Balsamic Vinegars

Paolo Giudici • Federico Lemmetti •
Stefano Mazza

Balsamic Vinegars

Tradition, Technology, Trade

Springer

Paolo Giudici
Department of Life Sciences
University of Modena and
Reggio Emilia
Reggio Emilia, Italy

Federico Lemmetti
Department of Life Sciences
University of Modena and
Reggio Emilia
Reggio Emilia, Italy

Stefano Mazza
Modena, Italy

ISBN 978-3-319-13757-5 ISBN 978-3-319-13758-2 (eBook)
DOI 10.1007/978-3-319-13758-2

Library of Congress Control Number: 2015931542

Springer Cham Heidelberg New York Dordrecht London

Printed on acid-free paper

Springer International Publishing AG Switzerland is part of Springer Science+Business Media (www.springer.com)

List of Abbreviations

AAB	Acetic acid bacteria
AGE	Advanced glycation end products
BC	Balsamic condiment
BI	Browning Index
BPF	Brown pigment formation
BV	Generic balsamic vinegar
BVM	Balsamic Vinegar of Modena
BVR	Base vinegar
BW	Base wine
CGM	Concentrated grape must
CI	Colour intensity
CKM	Cooked must
DOP	Denominazione di Origine Protetta
FA	Fixed acidity
GI	Geographical indication
GM	Grape must
GRAS	Generally recognised as safe
HMF	Hydroxymethylfurfural
HMW	High molecular weight
HORECA	Hotellerie Restaurant Cafè
IGP	Indicazione Geografica Protetta
IR	Infrared
LAB	Lactic acid bacteria
MEI	Methyl imidazole
PDO	Protected Denomination of Origin
PGI	Protected Geographical Indication
RD	Relative density
RT	Residence time
SEC	Size exclusion chromatography
TA	Titratable acidity
TBV	Traditional balsamic vinegar

TBVMO	Traditional Balsamic Vinegar of Modena
TBVRE	Traditional Balsamic Vinegar of Reggio Emilia
TSG	Traditional Speciality Guaranteed
VA	Volatile acidity

Contents

Chapter 1
Vinegar: Definition, Diffusion, and Uses

The history of microbial biotransformation is closely associated with vinegar production, which dates back to around 2000 years BC. However, among fermented foods, vinegar has always been considered a poor commodity because it is not a "food", lacks significant nutritional value, and it is produced from the transformation of richer and more nutritive fermented foods, such as wine and honey. Vinegar is a flavouring agent but also a preservative, and in some countries (Japan, China, and Korea) it is considered a healthy drink.

Vinegar production is a two step fermentation process, first by yeast and then by acetic acid bacteria, starting from almost any fermentable carbohydrate source: apples, pears, grapes, honey, syrups, cereals, hydrolysed starches, beer, and wine. With a few exceptions only, vinegar is an inexpensive commodity since low-cost raw materials are generally used for its production, like, for example, sub-standard fruits, seasonal agricultural surpluses, by-products from food processing, and fruit wastes.

Spirit vinegars are obtained directly through acetic oxidation of the ethanol from the distillation of fermented pulps and in some countries from petrochemical ethanol. Pyroligneous liquor, collected during wood carbonisation, is also called "vinegar" and it is used as an agricultural raw material, animal health product, cosmetic ingredient, and traditional medicine in Japan and East Asia (Mu et al. 2003, 2006).

Another subset includes flavoured vinegars, also known as herbal or fruit vinegars. Herbal vinegars consist of wine vinegars or white distilled vinegars, flavoured with garlic, basil, tarragon, cinnamon, cloves, nutmeg, or other herbs. Fruit-flavoured vinegars are wine and white vinegars sweetened with fruit or fruit juice to produce a characteristic sweet and sour taste. In these cases the fruit name does not indicate the raw material used for vinegar fermentation but rather the flavour added to achieve a specific taste and other characteristic features.

Some vinegars are very expensive and they are produced from regional and local foodstuffs according to longstanding traditions, such as Traditional Balsamic

P. Giudici et al., *Balsamic Vinegars*, DOI 10.1007/978-3-319-13758-2_1

Vinegar of Modena from Italy, Sherry vinegar from Spain, Oxos from Greece, and Shanxi mature vinegar from China.

1.1 Legal Definition and Composition

Vinegar production is regulated by an extensive set of statutes and the definition of vinegar varies from a country to another. The FAO/WHO defines vinegar as any liquid, fit for human consumption, produced exclusively from suitable products containing starch and/or sugars by the process of double fermentation, first alcoholic and then acetous. The residual ethanol content must be less than 0.5 % in wine vinegar and less than 1 % in other vinegars. (Joint FAO/WHO Food Standards Programme). In the USA, the Food and Drug Administration (FDA) requires that vinegar products must have a minimum acidity of 4 g per 100 ml, while in Australia the lower limit is 4 g per 100 g (FSANZ 2.10.1). This qualification ensures a minimum strength for vinegars sold in retail. There are currently no standards that identity vinegar; however, the FDA has established "Compliance Policy Guides" that the Agency applies for the labelling of vinegars, such as cider, wine, malt, sugar, spirit, and vinegar blends (FDA/ORA CPG 7109.22).

European countries have regional standards for vinegar produced or sold in specific areas. Unlike US law, the EU has established a threshold both for acidity and ethanol content. "Vinegar of X" is a general definition used for products with a minimum of 5 % (w/v) of acidity and a maximum of 0.5 % (v/v) of ethanol. Wine vinegar is obtained exclusively by acetous fermentation of wine, and it has at least 6 % of acidity (w/v) and 1.5 % (v/v) of ethanol (Regulation EC 479/2008). Table 1.1 shows the ranges of acidity and ethanol in some common vinegars.

Table 1.1 Acidity and residual ethanol content in various types of vinegars

Vinegar	Acidity (% w/v)	Ethanol (% v/v)
Malt vinegar	4.3–5.9	–
Cider vinegar	3.9–9.0	0.03
Wine vinegar	4.4–7.4	0.05–0.3
(semi-continuous process)	(8–14)	–
Rice vinegar	4.2–4.5	0.68
Chinese rice vinegar	6.8–10.9	–
Cashew vinegar	4.62	0.13
Coconut water vinegar	8.28	0.42
Mango vinegar	4.92	0.35
Sherry vinegar	7.0	–
Pineapple vinegar	5.34	0.67
Balsamic vinegar of Modena	>6.00	<1.5
Spirit vinegar		<20

In Europe vinegar is considered a flavouring or preservative food ingredient and, with few exceptions, its taste is generally sharp and sour. On the other hand, in Asia and Africa, vinegar is also a drink with a less sour taste. A lot of sweetened fruit vinegars characterised by low acidity and aromatic flavours are very popular in China and East-Southeast Asia. In Africa some fermented beverages can spontaneously acidify to produce alcoholic-acetous products, which are very difficult to classify as either alcoholic beverages or vinegars. Similarly, in Japan, black rice vinegar is usually diluted with fruit juice and consumed daily as a healthy tonic drink, representing 20 % of the Japanese vinegar market with a value of 21.46 billion yen in the year 2004. In some western countries, mainly the USA and Canada, apple cider vinegar is an old folk remedy claimed to be beneficial for treating a long list of diseases and it is consumed mixed with fruit juice. In the traditional wine-producing countries of Europe, it is very easy to differentiate between wine and vinegar, since the respective names are well established both historically and legally, with a precise definition for each. Wine can have a maximum of 1.2 g/l of acetic acid and, in all cases, the acetic acid content must be less than 1 % of the ethanol content. Vinegar must have a minimum of 6 % of titratable acidity and less than 1.5 % (v/v) of residual ethanol. In China, the word "vinegar" indicates products obtained from both fermentation and artificial processes, according to the Chinese National Standard definitions (CNS 14834, N5239 2004). In the previous National Industrial Standard of Vinegar, vinegar was classified in three grades, depending on its acetic acid concentration (3.5–4.5 %, 4.5–6 %, and >6 %, respectively). More recently, the Chinese State Administration Bureau for Quality and Technology issued a New National Standard Code of Condiments, introducing a definition of vinegar as either brewed or artificial (acetic acid blended with other ingredients, such as flavours). Each major vinegar also has its own local quality criteria and grading system. In the EU, some vinegars have a specific geographical indication (GI), a sort of intellectual property to protect regional food varieties and specialties, implemented by the European Union (EU) in a legal framework with three distinctive classes: Protected Designation of Origin (PDO), Protected Geographical Indication (PGI), and Traditional Specialty Guaranteed (TSG). See Chap. 8 for more details. Only a few vinegars bear a GI, an example being the Chinese Zhenjiang Xiang Cu (Table 1.2).

In general, vinegars are properly ascribed to the wider category of condiments, and they fall into two different sub-categories: with and without GI. Regarding the subset of balsamic vinegars, the main topic of this book, the grouping is more complex for at least two reasons: firstly, there is no precise and legally recognised definition of "balsamic vinegar"; and secondly, products bearing the "balsamic" tag can be very different one from each other (see Chap. 3).

Despite the fact that at first glance balsamic products can appear similar in appearance and sensory traits, they can actually differ markedly in ingredients, market claims, price, and legal status.

There are balsamic vinegars with more than 6 % of titratable acidity (TA, expressed as grams of acetic acid in 100 ml) and others with TA less than 4 %. TA is an important characteristic of vinegars. In oenology, TA is conventionally referred to the amount of strong basic molar solution necessary to neutralise (at pH

Table 1.2 Vinegars with geographical indications (GI)

Dossier number	Country	Name	App type	Date	Status
IT/PDO/ 0017/1565	Italy	Aceto Balsamico Tradizionale di Modena (ABTM)	PDO	20/04/ 2000	Registered
IT/PDO/ 0017/1566	Italy	Aceto Balsamico Tradizionale di Reggio Emilia (ABTRE)	PDO	20/04/ 2000	Registered
IT/PGI/ 0005/0430	Italy	Aceto Balsamico di Modena (ABM)	PGI	04/07/ 2009	Registered
ES/PDO/ 0005/0723	Spain	Vinagre de Jerez	PDO	05/10/ 2011	Registered
ES/PDO/ 0005/0724	Spain	Vinagre del Condado de Huelva	PDO	05/10/ 2011	Registered
ES/PDO/ 0005/0726	Spain	Vinagre de Montilla-Moriles	PDO	09/10/ 2012	Published
CN/PGI/ 0005/0630	China	镇江香醋 Zhenjiang Xiang Cu	PGI	14/06/ 2012	Registered

Source: http://ec.europa.eu/agriculture/quality/door/list.html

8.2) a specific amount of wine, and it is expressed in grams of tartaric acid per litre of wine.

The major contribution to wine TA arises from non-volatile organic acids, a fraction conventionally defined as *fixed acidity* that includes tartaric, malic, succinic, and lactic acids. The *volatile acidity* fraction consists almost entirely of acetic acid, and it should be less than 1 g/l. As for wine, the TA of vinegars is determined by titration, and it is conventionally expressed as acetic acid, since this acid is the principal organic component of most vinegars. The exception to this rule is the balsamic category, in which sugars are the major constituents, and fixed organic acids can be present in large amounts (see Chap. 3).

From a sensory point of view, fixed and volatile acids play important but different roles. Olfactory pungency is due to acetic acid, while fixed acids are detected only by taste (see Chap. 7). With regard to the acidity fraction, some balsamic products are very unusual because they do not have significant amounts of acetic acid or even TA. In the authors' opinion, and also in the legislation of several countries, this class of balsamic products cannot be defined as vinegar, belonging instead to the wider and more general category of condiments (see Chap. 3). This applies to the two PDO Traditional Balsamic Vinegars of Modena and Reggio Emilia (TBVs), for which the lower limit of TA is 4 % (w/w) and 4.5 % (w/v), respectively. In Italy, the minimum legal TA content for wine vinegar is 6 % (w/v), and furthermore the TA of the finest TBV consists mostly of fixed acids (Lemmetti and Giudici 2011).

From a commercial point of view, the official annual production of TBV is so small that it can be considered a food curiosity, at less than 300,000 bottles of 100 ml capacity, a total of just 30,000 litres! In reality, the overall production including the artisanal producers outside the consortium is larger, but a comprehensive survey does not exist.

In contrast, the other balsamic condiments without any sort of GI are commercialised in huge volumes around the world. These non-GI balsamic condiments differ from vinegars for their low acidity and their intended use. Balsamic condiments often imitate the characteristics of other expensive balsamic products, in some cases with good results although the price and quality vary widely and depend on the kind of raw materials used. Since balsamic condiments can contain thickeners, preservatives, colourants, flavours, and any other type of additive (both artificial and natural), these products are often called sauce, glaze, jelly, or dressing (see Chap. 3).

On the basis of certain particular characteristics and the technology applied, Shanxi mature vinegar could be grouped with balsamic vinegars, since it is cooked, concentrated, and its dark colour results from heat treatment and ageing. This product is the only balsamic vinegar not produced from grapes. Strictly speaking, there are also apple based balsamic vinegars, but the commercial quantities are practically zero).

1.2 Raw Materials and Critical Production Point

The majority of vinegars are of vegetable origin with only two exceptions: whey and honey. Whey is the residual milk serum from cheese making. It is rich in lactose and/or in the corresponding hydrolysed sugars, galactose and glucose, depending on the cheese making technology used. Honey is very rich in sugars (70–80 % w/w), mostly sucrose, fructose, and glucose, the ratio of which is influenced by the botanic origin of the nectar collected by the bees.

Several botanical species can be used for vinegar production on the condition they satisfy two main basic requirements. Firstly they must be safe for human and animal consumption, and secondly they must provide a direct or indirect source of fermentable sugar. General grouping can be made on the basis of the chemical composition of the edible parts of the plant and their proneness to fermentation:

- High acidity and ease of fermentation. pH below 3.5, glucose, fructose, and sucrose as main constituents; e.g. grapes, apples, plums, etc.
- Moderate acidity and ease of fermentation. pH between 3.5 and 4.5; e.g. figs, dates.
- Low acidity and ease of fermentation. pH higher than 4.5; e.g. palm sap.
- Non-fermentable. Hydrolysis is required before fermentation; e.g. seed crops and bananas.

The chemical composition of the raw materials exerts a strong selective pressure, effectively determining the dominant species of microorganisms involved in acetification.

The two fundamental steps in vinegar production are the preparation of the raw materials and fermentation. The first stage embraces all the necessary operations to ensure the availability of fermentable sugar and protein in solution, including slicing and/or crushing to obtain the fruit juice, enzymatic digestion of starch (for cereals), and in some cases cooking and steaming. In general, fruits require less processing than seed crops. Conversely, seed crops are more easily stored and preserved, making their use independent of harvesting time.

Fruits are highly perishable due to their high water content, and they need to be processed very quickly, in some circumstances (high temperatures, use of damaged fruits, etc.) even on the same day as harvesting. The logical outcome of these differences is that seed crops are typically transported and stored at large factories for transformation into vinegar. Instead, fruits tend to be processed at smaller, less technological factories, close to the production area.

1.3 Market Trends

From an economic point of view, vinegar production is a small industry in the overall economies of industrialised countries (Adams 1998). Global vinegar market shares in 2005 were balsamic vinegar (34 %), red wine vinegar (17 %), cider vinegar (7 %), rice vinegar (4 %), white vinegar (2 %), and finally other vinegars (36 %).

On the US market, white distilled vinegar has a 68 % unit share, cider vinegar 20 %, and specialty vinegar 12 % (data for 2007, www.versatilevinegar.org). In the specialty vinegar category, 39 % is from red wine, 30 % balsamic, 13 % all other wines, 12 % rice vinegar, and 6 % all other specialties.

The quantity of balsamic vinegar with GI, mostly Balsamic Vinegar of Modena (BVM), is in the range of 86–90 million litres (Anonymous 2014).

The main vinegar-producing countries are France, Italy, and Spain. In China the annual production for 2001 was 2.0×10^9 kg of brewed vinegar and 8.0×10^9 kg of distilled spirit vinegar (Wei 2001). In addition to white vinegar and fruit vinegars, brewed vinegar is also quite popular on the Chinese market. There are at least 14 types of traditional brewed vinegars, the six most widespread being Zhenjiang aromatic vinegar, Sichuan Bran vinegar, Shanghai rice vinegar, Jiangzhe rose vinegar, and Fujian red rice vinegar (Liu et al. 2004).

1.4 Use as a Preservative

Acetic acid is relatively weak and is well known as a very effective food preservative at low pH, and for this reason it is frequently associated with the strongest lactic acid. The bactericidal effect of acetic acid is related to the concentration of undissociated molecules in the medium. The non-ionic form of the acid is the only species able to cross the cellular membrane and enter the cytoplasm, where the pH is close to seven. Here the molecules release H^+ which dissipates the proton motive force of the cell (Eklund 1983, 1985; Salmond et al. 1984; Cherrington et al. 1990, 1991; Davidson 2001).

Vinegars with high TA are particularly effective against microorganisms and are used for the preparation of pickles, marinades, and sauces. In developing countries where food preservation technologies are scarce, vinegars can represent a very important option for preserving fresh fruit and vegetables from rapid deterioration, especially in the tropics where the prevailing environmental conditions accelerate food spoilage. Developing and improving small-scale vinegar production and food fermentation technologies in general was a goal of some FAO initiatives (Anonymous 1995).

Although the recipes for the preparation of pickles are numerous and often very elaborate, they all involve acidification, cooking, or pasteurisation, and then packaging which may follow or precede the heat treatment. The final pH value of the pickles must always be lower than 4.6, with at least 2 % of acetic acid in the brine. To achieve the required pH, a considerable amount of vinegar must be added to the vegetable raw materials, with a consequent marked modification of the sensory properties, and for this reason acidification of the brine is often achieved by addition of lactic and citric acid, which have a milder taste.

It is important to note that TBV's are not suitable for the preparation of pickles because of their low level of acetic acid, but they can be used as flavouring agents. Worldwide, the percentage of vinegar used as a preservative is close to 64 % of total vinegar production. It is used as an ingredient of dressings and sauces (16.8 %), for pickles (14.8 %), mustard (11.5 %), other processed foods (10.5 %), tomato products (8.5 %), and others (4.2 %). Source: http://www.versatilevinegar.org/markettrends.html.

1.5 Fermentation

Vinegar production involves different species of microorganisms during the various stages of the fermentation process, including lactic acid bacteria (LAB), yeasts, moulds, and acetic acid bacteria (AAB), whose habitats are often vegetable, fruit, and in general the raw materials used for vinegar production.

Two stages are common to all vinegars: alcoholic fermentation followed by acetic acid fermentation, respectively produced by yeasts and AAB. Other microorganisms are involved only in specific vinegars. Among yeasts, *Saccharomyces cerevisiae* is the most widespread species in fruit and vegetable vinegar, while the lactose fermenting yeast, *Kluyveromyces marxianus*, is the species responsible for whey fermentation. AAB are aerobic whole cell bio-catalysts involved in conversion of ethanol into acetic acid. AAB are common in fruits and in many sugar and acid environments, and their growth is promoted by procedures that increase oxygen availability after yeast fermentation. Finally, a physical association of yeasts, LAB, and ABB is involved in kombucha tea.

There are ten AAB genera (Yuzo and Pattaraporn 2008), but the majority of species detected in vinegars belong only to the Acetobacter and Gluconacetobacter genera (Gullo et al. 2006). However, this outline is not definitive because other species and genera involved in vinegar production are probably yet to be described. Furthermore, the AAB taxonomy is under extensive revision and the species and genera will soon be reorganised.

Fermentation can be induced either as spontaneous fermentation, by back-slopping, or by addition of starter cultures.

In spontaneous fermentation the raw materials are processed and the environmental condition selects the indigenous micro-flora. The more stringent the growth conditions, the higher the selective pressure exerted on the indigenous microorganisms. In very acidic and sugary environment, for example certain fruit juices, only yeast and AAB can grow.

Spontaneous fermentation is only suitable for small-scale production and very selective juices, because the method is difficult to control and there are risks of associated spoilage microorganisms. In the majority of spontaneous fermentations, microbial succession occurs, often with LAB and yeasts initially dominating. These microorganisms produce, respectively, lactic acid and ethanol, which inhibit the growth of many bacteria species, in addition to reducing sugar content and so prolonging the shelf life of the goods. Moulds mainly grow aerobically, and so their occurrence is limited to specific production stages or on crops before and after harvesting. Moulds are a major safety concern, since some genera and species produce aflatoxins. Therefore, the moulds used for starch hydrolysis of seed crops should be generally recognised as safe (GRAS) microorganisms.

Back-slopping involves the use of part of a previously fermented batch to inoculate a new batch. This procedure increases the initial number of helpful microorganisms and ensures more reliable and faster fermentation than would occur spontaneously. Back-slopping represents a crude form of starter culture, with the best adapted species seeded over the indigenous population (De Vuyst 2000). Nevertheless, the finished products are still exposed to the risk of fermentation failure, and moulds or harmful spoilage bacteria can develop.

In general, back-slopping is considered the most practical because it encourages the growth of helpful yeasts, inhibits the growth of spoilage and pathogenic microorganisms, and avoids the laborious and time consuming starter selection process.

Back-slopping is particularly useful to inoculate AAB, which are very fastidious microorganisms and it is very difficult to establish a true starter culture. In semi-continuous submerged acetification, at least one-third of vinegar volume is left in the fermenter to inoculate the new wine (Gullo and Giudici 2008; Gullo et al. 2009), whereas surface layer fermentation requires the physical transplantation of an AAB biofilm.

A starter culture can be defined as a microbial preparation comprising large numbers of cells of a microorganism (in some cases more than one), which is added to raw materials to produce a fermented foodstuff by accelerating and steering its fermentation process (Leroy and De Vuyst 2004).

Starter culture development is strictly related to the "pure culture" technique, a practice originally elaborated by R. Koch for bacteria. Using this approach, each microbial colony is the result of cells all deriving from the same original cell. This ensures that isolated cultures are not a mixture of different unknown individuals, and they can therefore be differentiated by specific traits, which form the basis for the systematic categorisation of microbes.

The use of starter cultures in food is a well consolidated practice, and it increases the safety, stability, and efficiency of the process, reducing production losses caused by uncontrolled fermentation and eliminating undesired properties. In oriental cereal vinegars, a mixed starter of unknown moulds and yeasts, called koji or Qu, respectively, in Japanese and Chinese, is used to hydrolyse and ferment rice and cereals. In other cases, starter cultures of oenological *S. cerevisiae* strains, selected for winemaking, are used to produce beer, wine, and cider, which become the alcoholic bases for vinegars. *S. cerevisiae* var. *sake*, selected for sake production, is used in rice vinegar fermentation.

The use of starter cultures in vinegar production is still far from being applied large scale for two main reasons: firstly, AAB are particularly fastidious microorganisms, and secondly, vinegars are generally inexpensive commodities produced by economical process.

References

Adams MR (1998) Vinegars. In: Wood BJB (ed) Microbiology of fermented food, vol 1. Blackie, London, pp 1–37

Anonymous (1995) Development education exchange paper. FAO, Rome

Anonymous (2014) http://www.ansa.it/terraegusto/notizie/rubriche/prodtipici/2013/12/09/Aceto-balsamico-Modena-costituito-unico-Consorzio-tutela-_9750815.html. Retrieved 18 Sept 2014

Cherrington CA, Hinton M, Chopra I (1990) Effect of short-chain organic acids on macromolecular synthesis in *Escherichia coli*. J Bacteriol 68:69–74

Cherrington CA, Hinton M, Mead GC et al (1991) Organic acids: chemistry, antibacterial activity and practical applications. Adv Microb Physiol 32:87–108

Chinese National Standard (2004) Edible vinegar. No.14834, N5239

Council Regulation (EC) No 479/2008 of 29 April 2008 on the common organisation of the market in wine, amending Regulations (EC) No 1493/1999, (EC) No 1782/2003, (EC) No 1290/2005, (EC) No 3/2008 and repealing Regulations (EEC) No 2392/86 and (EC) No 1493/1999

Davidson PM (2001) Chemical preservatives and natural antimicrobial compounds. In: Doyle MP, Beuchat LR, Montville TJ (eds) Food microbiology—fundamentals and frontiers, 2nd edn. American Society for Microbiology, Washington, DC, pp 593–627

De Vuyst L (2000) Technology aspects related to the application of functional starter cultures. Food Technol Biotechnol 38:105–112

Eklund T (1983) The antimicrobial effect of dissociated and undissociated sorbic acid at different pH levels. J Appl Bacteriol 54:383–389

Eklund T (1985) Inhibition of microbial growth at different pH levels by benzoic and propionic acids and esters of p-hydroxybenzoic acid. Int J Food Microbiol 2:159–167

Food and Drug Administration, FDA/ORA Compliance Policy Guides 7109.22 Sec 525.825. Vinegar, definitions—adulteration with vinegar eels
FSANZ (2008) Australia New Zealand Food Standards Code, Standard 2.10.1 Vinegar and related products
Gullo M, Giudici P (2008) Acetic acid bacteria in traditional balsamic vinegar: phenotypic traits relevant for starter cultures selection. Int J Food Microbiol 125:46–53
Gullo M, Caggia C, De Vero L, Giudici P (2006) Characterization of acetic acid bacteria in "Traditional Balsamic Vinegar". Int J Food Microbiol 106:209–212
Gullo M, De Vero L, Giudici P (2009) Succession of selected strains of Acetobacter pasteurianus and other acetic acid bacteria in Traditional Balsamic Vinegar. Appl Environ Microbiol 75:2585–2589
Lemmetti F, Giudici P (2011) Semplici parametri chimico-fisici per riconoscere gli aceti balsamici invecchiati. Industrie Bevande 40:12–21
Leroy F, De Vuyst L (2004) Lactic acid bacteria as functional starter cultures for the food fermentation industry. Trends Food Sci Technol 15:67–78
Liu D, Zhu Y, Beeftink R et al (2004) Chinese vinegar and its solid-state fermentation process. Food Rev Int 20:407–424
Mu J, Uehara T, Furuno T (2003) Effect of bamboo vinegar on regulation of germination and radicle growth of seed plants. J Wood Sci 49:262–270
Mu J, Yu Z, Wu W, Wu Q (2006) Preliminary study of application effect of bamboo vinegar on vegetable growth. Forest Sci Pract (formerly Forestry Studies in China) 8:43–47
Regulation (EC) No 479/2008 on the common organisation of the market in wine as regards support programmes, trade with third countries, production potential and on controls in the wine sector. Offic J L 170(30/06/2008):0001–0080
Salmond CV, Kroll RG, Booth IR (1984) The effect of food preservatives on pH homeostasis in *Escherichia coli*. J Gen Microbiol 130:2845–2850
The Vinegar Institute. http://www.versatilevinegar.org/markettrends.html. Accessed January 2014
Wei X (2001) A milestone of condiments administration with industrial standard. Chin Condiment 1:3–8
Yuzo Y, Pattaraporn Y (2008) Genera and species in acetic acid bacteria. Int J Food Microbiol 125 (1):15–24

Chapter 2
History of Balsamic Vinegars

2.1 Mystery and Myth: The Balsamic Family

Amongst the most famous condiments produced in Europe, and more precisely in Italy, there is a group of vinegars that have recently achieved outstanding world-wide fame for their peculiar historical *iter* and recent commercial success: the "balsamic vinegars".

The name "*balsamic vinegar*" is a translation of the Italian name *aceto balsamico*, which indicates a heterogeneous group of vinegars with some common features. The Italian word *balsamico* ("balsamic" in English) originates from an adjective that means soothing, healing and, more generally, good for health. It is used for certain balms, officinal plants or products from the same, and balsamic vinegars.

Balsamic vinegar is a traditional name that nowadays is used to indicate a heterogeneous group of condiments made from grape must, originating from Italy. The market for balsamic vinegars and related products has developed very quickly over a short period of time, embracing a wide range of products which at first impression seem quite similar in appearance and even from the sensory point of view, but these products can differ widely as regards ingredients, market claims, price, and legal status (see Chap. 3).

The various forms of balsamic vinegar are traditionally produced in two towns in northern Italy: Modena and Reggio Emilia. Produced according to the traditions of their original territories, balsamic vinegars are characterised by a strong local identity, as well as chemical–physical and sensory properties defined by international regulations with Protected Geographical Indication for the Balsamic Vinegar of Modena (*Aceto Balsamico di Modena* I.G.P.), and Protected Denomination of Origin for the Traditional Balsamic Vinegar of Modena and Traditional Balsamic Vinegar of Reggio Emilia (*Aceto Balsamico Tradizionale di Modena* D.O.P and *Aceto Balsamico Tradizionale di Reggio Emilia* D.O.P.).

P. Giudici et al., *Balsamic Vinegars*, DOI 10.1007/978-3-319-13758-2_2

The main features of balsamic vinegars are their pungent, sweet taste, dark colour, and complex sensory profile, along with the typical maturation process they undergo. Notwithstanding the claims of a long, time-honoured tradition, the culture, history and complexity behind these products are far from being correctly reconstructed and properly understood. No serious and comprehensive research into the original production process, maturation system, sensory profile, and the analytical parameters that fully describe balsamic vinegars has been conducted to date.

The remainder of this book will discuss and examine the legal definition of the vinegars and condiments in depth. For the purpose of the present chapter, they will be referred to simply as "balsamic vinegars", not out of disrespect for the official names they have been legally awarded today, but to underline their undefined nature prior to modern times. The aim is to illustrate as precisely as possible the long and complex history of balsamic vinegar, from its most remote ancient origins to the present day. Obviously, it is impossible to draw a complete and accurate picture, but at least all the sources available can be examined and integrated together.

Despite their well consolidated presence in the culinary habits of the entire world, balsamic vinegars still remain something of a mystery. Two specific mysteries surround these famous vinegars, neither of which has been successfully resolved to date.

The first is their origin since nobody knows with certainty how these vinegars were invented. Balsamic vinegars are generally described as being of ancient origin, dating back to the early Middle Ages, and probably even earlier, and as is well known, they are deeply rooted in the culinary history of the Italian provinces of Modena and Reggio Emilia. Beyond this long established tradition, the culture and complexity of their historical evolution are, however, far from being understood.

The second mystery surrounding balsamic vinegars, no less intriguing, is their method of production. Among the many sources available, which will be discussed in the following paragraphs, it clearly emerges that it is impossible to establish a single form of production definable as the genuine original, also because the members of the balsamic family are very different from each other and were developed in different periods.

Fortunately, not everything is grey and undefined, and there are certain key events in the history of balsamics that stand out as landmarks that remain to the present day, forming the main subject of this chapter.

It appears reasonably certain that the current manufacturing techniques derive from practices developed in the remote past. From a regulatory point of view, things may seem simple, especially after international protection was officially granted with Protected Denomination of Origin (PDO) for the Traditional Balsamic Vinegar of Modena and Reggio Emilia in the year 2000 and Protected Geographical Indication (PGI) for the Balsamic Vinegar of Modena in the year 2009. However, while the Regulations comprehensively specify parameters, times, and raw materials, they do not define a single production process valid for all producers, who remain suspended in a scenario without clear reference points. Thus, a combination of experience, techniques, and personal decisions of manufacturers according to their own methods generates the incredible differentiation of sensory characteristics observed in balsamic vinegars. Anybody who wants to understand this product still has to rely on the traditions passed down through families in the cities of Modena and Reggio Emilia, passing on from them to the large scale producers and companies that supply the vast world market today. This aspect undoubtedly enhances the prestige of balsamic vinegars, giving the general public a marked impression of originality, authenticity, and uniqueness, which would be impossible to achieve with a more technical methodology.

The aim of this chapter is to bring to light as much evidence as possible and initiate a reliable, documented study of the complex history of the entire *balsamic family*.

2.2 Balsamic, a Linguistic Overview

The word *balsamic* is attested in the English language since the Middle Ages as an adjective formed from the noun *balsam*, which means "any substance able to gives, restore, or comfort," or alternatively, "certain officinal plants and/or products thereof". The words *balsamic* and *balsam* have passed through different language families over a very long period of time. In modern European and American languages, the linguistic forms of the noun and derived adjective are well conserved, thanks to the shared Latin and Greek origin, as can be seen in Table 2.1.

The various forms of the word *balsam* all originated from the Latin *balsamum*, which in turn was derived from the Ancient Greek *bálsamon*, attested since the fourth century BCE.

The form *bálsamon* was borrowed from the Hebrew *baśam*, related to Aramaic *busma* and the Arabic *baśam*, meaning "balsam, spice, perfume, incense", as seen in Table 2.2.

Table 2.1 Comparison of the word "balsam" in different European languages

Language	Form	Attestation dates
Modern English	balsam, balm	After 1500 AD
Old Spanish (Old Castilian)	balsamo (?)	900–1500 AD
Middle English	balsamum/balsaum basme	1175–1225 AD 1220 AD
Old French	basme/balsme/balme	1000–1300 AD
Old Norse	balsam	1000–1300 AD
Old English	balsam/balsame	Before 1000 AD
Old High German	balsamo	Around 1000 AD
Vulgar Latin (*Toscano* Italian)	balsamo/balsimo	Since 900 AD

Data retrieved from The American Heritage Dictionary of the English Language (2004); The Merriam-Webster Online Dictionary; Köbler (2006); Vocabolario degli Accademici della Crusca on-line

Table 2.2 The borrowing of the forms of *balsam* by Latin and Ancient Greek from Semitic languages

Language	Form	Attestation dates
Latin	balsamum	4th century BCE
Ancient Greek	bálsamon	4th century BCE
Hebrew	bośem/baśam	
Aramaic	besma/busma	
Arabic	baśam	

Data retrieved from The Merriam-Webster Online Dictionary; The American Heritage Dictionary of the English Language (2004)

The root from which the word *balsam*/*balsamic* formed is thus clearly of Semitic origin and is usually represented as *bśm* (The American Heritage Dictionary of the English Language, Appendix II: "Semitic roots", 2000). As regards the ancestry of this root, very little is known and it can only be hypothesised that the Semitic *bśm* comes from an even older root present in a larger number of ancient languages (Fig. 2.1).

2.3 From the Remote Past: The Origins (510 BCE–530 CE)

The search for reliable information from the available sources and subsequent reconstruction of the true story behind balsamic vinegars is a very challenging task, because the documentation and testimonies of them are few and sometimes misleading. However, it is well established in popular tradition that the condiments renowned throughout the world as balsamic vinegars originated in the Emilia region of Italy, and as far as can be deduced from more recent history, they have been continuously produced in the form that we know today in the two provinces of Modena and Reggio Emilia (Mazza and Muruoka 2008; Rangone and Giacobazzi 2011).

It is very likely that something similar to these condiments was already used in Ancient Roman times. The remaining historical sources relate that in Roman times, it was common to boil soured wine in lead pots[1] to obtain a very thick, sweet syrup called *sapa* in Latin. This is probably the origin of modern Italian *saba*, which is an alternative name for cooked grape must. Cooked must is mentioned quite often in Roman literature, examples including the works of popular authors like Cato (Marcus Porcius Cato, 234 BCE–149 BCE), Varro (Marcus Terentius Varro, 116 BCE–27 BCE), Columella (Lucius Junius Moderatus Columella, 4 CE—approx. 70 CE), and Pliny the Elder (Gaius Plinius Secundus, 23 CE–79 CE). The uses of cooked grape must appear to have been rather varied, including the

[1] Operation that must not be replicated for any reason: it results in a solution of lead acetate that is a lethal compound.

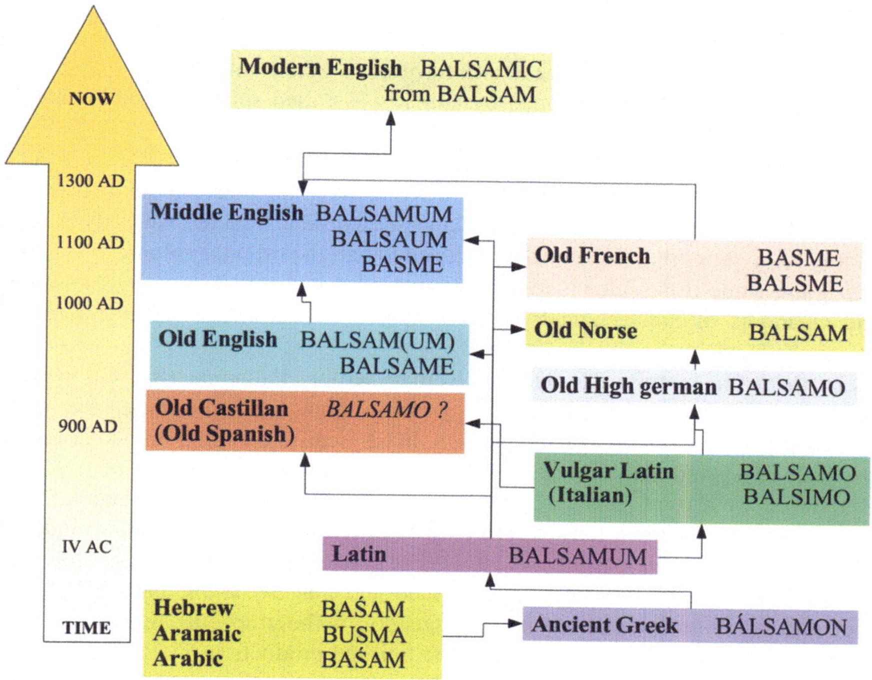

Fig. 2.1 Evolutionary timeline of the term *balsamic*

preservation of fruits like grapes, pears, sorbs, cornelian cherries,[2] plums, and olives. It was also used to improve the taste of sour wine or wine with other defects and for preparing the "light wine" consumed by slaves. It was also recommended for medical applications. The ancient authors refer to the cooked must with two different terms: *defrutum* and *sapa*. The two products differ mainly in cooking time during preparation. Cato and Varro do not give specific indications for preparation, but Columella and Pliny the Elder provide more details.

In the first century CE, Columella wrote a famous agricultural treatise called *De Re Rustica*. In the first book of the treatise, he indicates that the grape juice must be boiled in a place that "should be neither cramped nor dark", so that the servant in charge of cooking has enough room to move around, and that the firewood for cooking should be prepared before the harvest. Two chapters of the twelfth book are devoted entirely to the preparation of cooked grape must: according to the method described, the cooking vats must be lead and not bronze, they must be greased with oil before pouring the liquid inside, and the grape juice must be boiled slowly over a small fire. The deposits that form on the bottom during cooking should be removed using fennel sheaves tied onto a wooden stick, while the foam that forms on the

[2] The fruit of *Cornus mas*, also called Cornelian cherry, European cornel or dogwood.

surface can be removed with drainers made of soft rush or esparto. Finally, when cooking was almost complete, various flavourings could be added. If the final cooked grape must was reduced to a third of the original volume, the product was called *defrutum*, if it was reduced by half it was called *sapa* (Columella, *De Re Rustica*).

Pliny the Elder does not provide as detailed information as Columella, and in his book *Naturalis Historia* he simply specifies that the grape must should be boiled in lead or bronze containers and that cooking should be conducted when there is no moon. Pliny states that if the juice is reduced to half the original volume, it is called *defrutum*, while if the juice is reduced to one-third it is called *sapa*, thus inverting the meanings of the technical terms compared to Columella (Pliny the Elder, *Naturalis historia*).

In the fifth century CE another Roman writer, Palladius (Rutilius Taurus Aemilianus Palladius, IV CE – ?), wrote a book on agriculture, the *Opus agriculturae*, also known as *De re rustica*. In his treatise on agronomy he discusses the topic of cooked must and defines three grades: *sapa* if reduced to half, *defrutum* if reduced to a third, and *caroenum* if reduced to two-thirds (Palladius, *Opus agricolturae*). The latter term is still used today in some parts of Italy to indicate a cooked syrup obtained from cooked wine or cooked grape must. It is obvious that the terminology was not rigorously defined and that *defrutum*, *sapa*, and, more recently, *caroenum* were used interchangeably, without specific reference to the cooking process or the final density of the finished products.

The cooked grape must, be it *sapa*, *defrutum*, or *caroenum*, was probably stored for long periods and consumed day to day, exposed to the air and environment. In those conditions, it could undergo spontaneous transformations and become a sort of vinegar itself or a condiment produced by transformations brought about by spontaneously selected microflora. Unfortunately, after Columella and the other Roman authors, there are no other written sources on this topic until the Middle Ages, when the first documents appeared expressly reporting a vinegar considered of such high value as to be a worthy gift for nobles and kings.

2.4 The Middle Ages (600 CE–1400 CE)

The Middle Ages was the period of European history from the fifth century to the fifteenth century following the fall of the Western Roman Empire. Many changes occurred across the whole continent during this period, including cultural aspects like food production and consumption. Once again, it is necessary to make use of written documents and testimonies to try and reconstruct the culture and usage of vinegar during this period. It appears that vinegar had become a widespread condiment, well known for its culinary and medical properties, and is frequently cited by numerous medieval authors in their writings. The first written treatise that mentions vinegar in general is probably *De rerum naturis* by Rabanus Maurus Magnentius, a Carolingian scholar and clergyman in Germany, who composed his book between AD 842 and 846 (Giudici and Solieri 2008). It was possibly in this historical period that balsamic vinegar started to assume its present form, with gradually increasing fame all over Europe. It is impossible to establish the exact location where this specific variant of the condiment was made for the first time, and it might have been Modena, Reggio Emilia, or another town in the Emilia region in Italy. However, the first mention of this valuable locally produced condiment is presumed to be in the poem *Acta Comitissae Mathildis*, written in the twelfth century by the monk Donizo of Canossa (eleventh to twelfth century) and later known as *Vita Mathildis*. Donizo recounts that in the year 1046, Count Bonifacio III, one of the most powerful northern Italian princes of the time, had his castle in Canossa, within the Duchy of Modena and Reggio Emilia. He presented vinegar as a gift to Henry III of Franconia (Donizo (2008); Book 1, Chapter 13). An important detail is that in the Italian poem the adjective "balsamic" is not used, even if the words "balsam" and "balsamic" were already present in the local Italian dialects ever since they split off from the original Latin language (see sect. 2.2 in this chapter). This ancient document simply mentions a famous vinegar made in this area, but nothing is said about its properties and composition. Similarly, other early sources mentioning vinegar are often cited in order to demonstrate the ancient tradition of "balsamic" vinegar making, but these documents from the area around Modena and Reggio Emilia always cite the word "vinegar" alone. However, even if the word "balsamic" is absent from the texts, it is a logical supposition that this vinegar must have been of very high quality to be considered a gift worthy of an emperor.

Vinegar became popular in this historical period, partly due to its dietary and medical properties, and not least because it was one of the safest substances known at the time. Many pharmacological and medical handbooks were compiled in Italy starting from the fourteenth century, generally called "health tables" or *Tacuina sanitatis* in Latin. These books described the various properties of foods and plants and gave suggestions on physical exercise, mental health, rest, and general guidelines for good health. Among these books, the *Tacuinum Sanitatis Casanatense*, written in Milan, Italy, is of particular interest because it includes a picture of a lady climbing a ladder and filling a pitcher from a cask of vinegar (see Fig. 2.2). An important detail in the picture is that the cask is kept near the roof of a small building, probably a stable or a shed. This illustration shows how domestic spaces for vinegar production and storage were organised, with the vinegar stored mainly up high, in attics or other raised parts of the house, a custom which remains strong to this day (Mazza and Muruoka 2008).

During the eighteenth century, the Italian physician and naturalist Antonio Vallisnieri records that several casks of vinegar were kept at the Court of Obizzo II of Este, in the year 1228, but once again it can only be inferred from the context that this was probably a valuable vinegar, without precise indications of its nature and whether or not it was *balsamic*.

2.5 The Renaissance and Enlightenment (1400–1700)

After the Middle Ages, there is extremely little useful documentation until the end of the sixteenth century, with only a handful of details, like for example the records of the Duke of Este in Modena for the year 1556, which exhibit an unusual

Fig. 2.2 Tacuinum sanitatis Casanatense: Vinegar. Rome, Biblioteca Casanatense

classification of stored vinegars, which were intended for separate uses (Saccani and Ferrari Amorotti 1999).

Later in the same century, some more useful documents appear, revealing increasing interest for vinegars and recipes for making different types. In 1570, Charles Estienne, a renowned French scientist and publisher, for the first time in history published a recipe for a so-called "sweet vinegar". The recipe is reported in his treatise *L'agriculture et maison rustique*, which was revised by Jean Liébault, published in Paris in the year 1570. In Book 5 of the treatise, Chapter 57, titled *Some peculiarities concerning vinegar*, Estienne gives specific instructions to make this "sweet vinegar" (*vinaigre doux*, in French) (Fig. 2.3):

> To make sweet vinegar: take five *sextiers*[3] of strong vinegar and the same amount of good grape must, which was stored separately after the grapes were pressed. Then add pitch[4] and

[3] It is assumed that the Old French word *sextier* in the text corresponds to the modern French *setier*, here possibly used as a synonym of *chopine*, a measurement unit that indicated approximately 0.45 L. Comparison was also made with the 1591 Italian translation by Hercole Cato, H., which indicated this as equivalent to "two or three *inghistare*" (Cato 1591). An *inghistara* was about 0.979 l (Zupko 1981).

[4] Pitch: it is not clear why the author suggests adding pitch to the vinegar mixture. One hypothesis is that it could be for colouring purposes, since the maturation period of 1 month suggested in the text is very short to achieve the typical dark colour of balsamic vinegar. If the pitch was tar obtained from wood or peat, it might also serve as a flavour enhancer or even to prevent undesired fermentation, but the latter hypothesis appears the weakest.

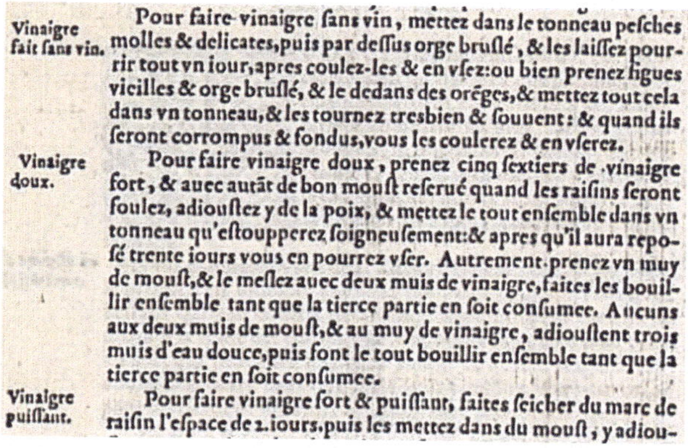
Vinaigre fait ſans vin. Pour faire vinaigre ſans vin, mettez dans le tonneau peſches molles & delicates, puis par deſſus orge bruſlé, & les laiſſez pourrir tout vn iour, apres coulez-les & en vſez: ou bien prenez figues vieilles & orge bruſlé, & le dedans des oréges, & mettez tout cela dans vn tonneau, & les tournez tresbien & ſouuent: & quand ils ſeront corrompus & fondus, vous les coulerez & en vſerez.

Vinaigre doux. Pour faire vinaigre doux, prenez cinq ſextiers de vinaigre fort, & auec autãt de bon mouſt reſerué quand les raiſins ſeront foulez, adiouſtez y de la poix, & mettez le tout enſemble dans vn tonneau qu'eſtoupperez ſoigneuſement: & apres qu'il aura repoſé trente iours vous en pourrez vſer. Autrement prenez vn muy de mouſt, & le meſlez auec deux muis de vinaigre, faites les bouillir enſemble tant que la tierce partie en ſoit conſumee. Aucuns aux deux muis de mouſt, & au muy de vinaigre, adiouſtent trois muis d'eau douce, puis font le tout bouillir enſemble tant que la tierce partie en ſoit conſumee.

Vinaigre puiſſant. Pour faire vinaigre fort & puiſſant, faites ſeicher du marc de raiſin l'eſpace de 2. iours, puis les mettez dans du mouſt, y adiou-

Fig. 2.3 A detail from the treatise *L'agriculture et maison rustique* by Charles Estienne and Jean Liébault, first edition 1570, which contains a recipe to make a "sweet vinegar" (*vinaigre doux* in French). The excerpt is taken from Book 5, Chapter 57, titled *Some peculiarities concerning vinegar*

> put all the mixture together in a barrel, which must be carefully sealed. After thirty days maturation it is ready for consumption. As an alternative method, take a barrel of must and mix it with two barrels of vinegar, boil them together until one third is evaporated. Some people add three barrels of fresh water to two barrels of must and two barrels of vinegar, then boil the mixture until one third of the liquid has evaporated (Estienne and Liébault 1570).

An interesting detail is that Charles Estienne and Jean Liébault were *French* scholars, rather than Italian. They had some connections with Italy, in particular Estienne, who spent part of his life in Italy, in the cities of Venice and Genoa, but there is no record of him visiting the city of Modena, or the Emilia region. On the basis of historical facts alone, the only observation is that two Frenchmen in the middle of sixteenth century knew at least two methods for making a "sweet vinegar", obtained basically by blending grape must with wine vinegar and then maturing it in wooden casks. The second recipe in particular clearly indicates a reduction method by heating, which is very similar to the one used on a larger scale to this day in vinegar factories to prepare various types of balsamic vinegar.

Another minor but interesting document dates to the year 1597. It is a letter written by the Duke's Attorney, referred to as G. F. Vezzali, and addressed to the general farm manager Ercole Estense Mosti, commissioning the purchase of *Trebbiano* grapes for the *accete*[5] (Saccani and Ferrari Amorotti 1999).

One year later, in 1598, Modena became the capital city of the Duchy of Modena and Reggio Emilia under the domain of Cesare d'Este and his family. When the Este family moved to Modena, they carried their possessions with them, but here they encountered a new kind of vinegar, unknown to most people, home made on a very small scale from time immemorial. It was an unusual vinegar with peculiar features including a unique and distinguished flavour. They immediately

[5] *Accete* is an ancient word, cognate of the Italian *aceto* (vinegar), and probably a borrowing from a local dialectal form, which could indicate both the wooden barrels in which the vinegar is produced and aged, or even the premises where the barrels are stored. The modern Italian word *acetaia* (a place, typically an attic or cellar, for the production of vinegar) derives from this term.

recognised it as an excellent product that stood above all more widely known varieties, and the Duke's palace soon had its own *acetaia* (Saccani and Ferrari Amorotti 1999). It was probably for this reason that this highly valued vinegar also came to be known as "the Duke's vinegar" (*l'aceto del Duca*).

2.6 The Casks and the Dukes (1764–1859)

It should be noted that, while a number of references to the vinegars produced in the areas of Modena and Reggio Emilia exist, no direct mention of the adjective *balsamic* (*balsamico* in Italian) has ever been found from this initial period. This makes the year 1747 particularly important in the present context, because there is the first documented record of the adjective balsamic in the inventories of the Este Court, mainly in records of donations or nuptial gifts. From this time onwards, the vinegar previously known as *the Duke's vinegar* started to be called balsamic (Saccani and Ferrari Amorotti 1999), possibly in order to distinguish it from other similar looking vinegars made with grape must, but of lesser value.

The Dukes of Modena proved to be very active promoters of balsamic vinegar among the noble families of Europe, regurarly giving their precious condiment to guests and rulers as a gift. In 1764 Francis III of Este gave some small bottles to the Count Mikhail Illarionovich Vorontsov, a Russian statesman on a diplomatic mission to Modena, for Catherine the Great. In 1792, Ercole III of Este sent a single bottle of balsamic vinegar in Frankfurt, to honour the coronation of Francis II of Austria as Holy Roman Emperor (Saccani and Ferrari Amorotti 1999).

The popularity of these kinds of vinegar was so great that Modena became internationally famous for them. At the end of eighteenth century, the Hungarian scientist and agronomist Ludwig (Ljudevit) Mitterpacher von Mitterburg wrote a book on agronomy and farm products entitled "Elements of Agriculture". It became very popular in his time, even in Italy, where it was published in the Italian language in 1784. This treatise dedicates an entire chapter to vinegar, describing many different methods for making it, and is very interesting because it demonstrates how vinegar making varied in different parts of Europe, using different approaches to achieve good products with the desired features, the most important of which was acidity (Fig. 2.4).

His exhaustive review of the various methods of production includes a paragraph dedicated to the vinegar of Modena (*alla modenese* in the text), with its very unusual form of preparation (Fig. 2.5):

> Also famous is the vinegar of Modena, which is made in the following way. The must of white grapes is boiled for three days together with the stalks inside a vat, then it is transferred to a separated container, leaving the stalks apart in the vat, where they warm up for three days. After this, they are blended again with the grape must and everything is left to lie for eight days. Then the must is divided into separate tubs, and a heated tile is put into each one. After that, the treated must is transferred into casks, where after one year it becomes vinegar, and can be transferred through the sequence of barrels (Mitterpacher 1784).

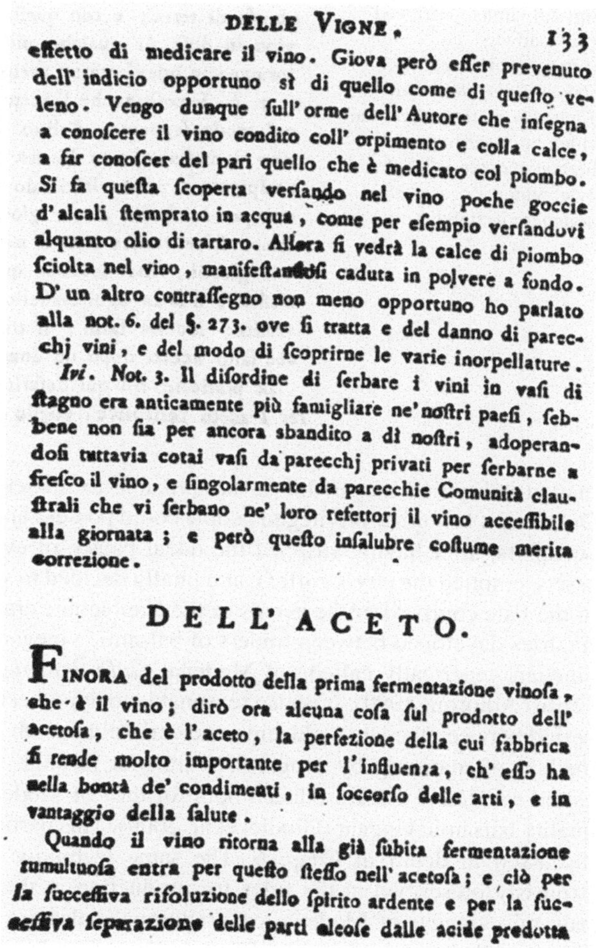

DELLE VIGNE. 133

effetto di medicare il vino. Giova però eſſer prevenuto dell' indicio opportuno sì di quello come di queſto veleno. Vengo dunque ſull' orme dell' Autore che inſegna a conoſcere il vino condito coll' orpimento e colla calce, a far conoſcer del pari quello che è medicato col piombo. Si fa queſta ſcoperta verſando nel vino poche goccie d' alcali ſtemprato in acqua, come per eſempio verſandovi alquanto olio di tartaro. Allora ſi vedrà la calce di piombo ſciolta nel vino, manifeſtandoſi caduta in polvere a fondo. D' un altro contraſſegno non meno opportuno ho parlato alla not. 6. del §. 273. ove ſi tratta e del danno di parecchj vini, e del modo di ſcoprirne le varie inorpellature.

Ivi. Not. 3. Il diſordine di ſerbare i vini in vaſi di ſtagno era anticamente più famigliare ne' noſtri paeſi, ſebbene non ſia per ancora sbandito a dì noſtri, adoperandoſi tuttavia cotai vaſi da parecchj privati per ſerbarne a freſco il vino, e ſingolarmente da parecchie Comunità clauſtrali che vi ſerbano ne' loro refettorj il vino acceſſibile alla giornata; e però queſto inſalubre coſtume merita correzione.

DELL' ACETO.

FINORA del prodotto della prima fermentazione vinoſa, che è il vino; dirò ora alcuna coſa ſul prodotto dell' acetoſa, che è l' aceto, la perfezione della cui fabbrica ſi rende molto importante per l' influenza, ch' eſſo ha nella bontà de' condimenti, in ſoccorſo delle arti, e in vantaggio della ſalute.

Quando il vino ritorna alla già ſubita fermentazione tumultuoſa entra per queſto ſteſſo nell' acetoſa; e ciò per la ſucceſſiva riſoluzione dello ſpirito ardente e per la ſucceſſiva ſeparazione delle parti alcoſe dalle acide prodotta

Fig. 2.4 Image from the chapter *Dell'Aceto* ("Concerning Vinegar") from the book *Elements of agriculture* by Ludwig Mitterpacher (1784), Italian edition

In 1796, Ercole III of Este, Duke of Modena and Reggio Emilia, left his normal abode after the occupation of the city of Piacenza by French troops under Napoleon and took refuge in the safer city of Venice. It is reported that he took his casks of vinegar with him, and various other reported details make it clear that the commercial value of this vinegar was reasonably high, even if many variants of differing quality already existed There are widespread references to the difference between a quality defined as *common* (*comune*), which included mostly wine vinegar, and a more distinguished quality defined in Italian as *da padrone*, which means *master's* or *noble*, and which was produced starting from grape must (Rangone and Giacobazzi 2011).

This period presents another interesting historical fact: until the Napoleonic invasion of the Duke's territory in 1796, nobody had ever thought or dared to

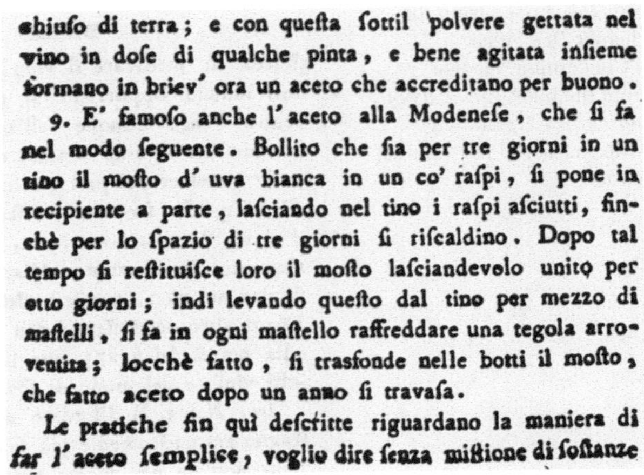
chiuſo di terra; e con queſta ſottil polvere gettata nel vino in doſe di qualche pinta, e bene agitata inſieme formano in briev' ora un aceto che accreditano per buono.

9. E' famoſo anche l'aceto alla Modeneſe, che ſi fa nel modo ſeguente. Bollito che ſia per tre giorni in un tino il moſto d' uva bianca in un co' raſpi, ſi pone in recipiente a parte, laſciando nel tino i raſpi aſciutti, finchè per lo ſpazio di tre giorni ſi riſcaldino. Dopo tal tempo ſi reſtituiſce loro il moſto laſciandevolo unito per otto giorni; indi levando queſto dal tino per mezzo di maſtelli, ſi fa in ogni maſtello raffreddare una tegola arroventita; locchè fatto, ſi traſfonde nelle botti il moſto, che fatto aceto dopo un anno ſi travaſa.

Le pratiche fin quì deſcritte riguardano la maniera di far l'aceto ſemplice, voglio dire ſenza miſtione di ſoſtanze

Fig. 2.5 Image retrieved from Ludwig Mitterpacher's treatise *Elementi d'agricoltura (Elements of agriculture)* in which he describes a method to make a "Modena-style" vinegar (*aceto alla modenese* in Italian)

market balsamic vinegar. It was unthinkable to monetise a product that only the Duke and a few other privileged people could possess and use. However, the French occupiers, after having stripped the ducal palace of everything, sold the national assets, emptied the city's coffers, and finally decided to sell off the balsamic vinegar of the Este court. After the events of the Napoleonic era in Italy, many commercial rivalries developed between traders of balsamic vinegar and various other kinds of vinegar, generically called "of Modena" (*alla modenese* in Italian) (Saccani and Ferrari Amorotti 1999). It is also reasonable to hypothesise that during the historical period between the late eighteenth and early nineteenth centuries, the characteristic method of maturing by progressive transfer through barrels of decreasing size, called *rincalzo*,[6] was introduced both for the "of Modena" vinegar and the higher quality balsamic vinegar (Giudici et al. 2009). This particular method of maturing is discussed in detail in Chap. 6. The same technique is still used today and is expressly prescribed in the rules for production in the two PDO's for traditional balsamic vinegar of Modena and traditional balsamic vinegar of Reggio Emilia. The so-called*rincalzo* involves fractional blending in a sequence of barrels, so that the finished product is a mixture of ages, with the average age gradually increasing as the process continues over many years. The vinegar is refined in barrel sets of different sizes and wood varieties. Every year, a small quantity of the finished matured vinegar is tapped off from the smallest barrel. This barrel is then topped up from the content of the preceding barrel with the operation repeated backwards to the first and largest cask, which then receives the newest cooked grape must (Giudici and Rinaldi 2007).

This kind of maturing is not exclusive to traditional balsamic vinegars, but it is an established part of their culture and mandatory in the production regulations. It is possible that the *rincalzo* method did not originate in Emilia or of Italy. It is clearly

[6] An English translation of *rincalzo* could be *topping up* or *back-filling*.

a system invented to save time from year to year in the maturing of each barrel and is virtually identical to the Spanish *solera* wine system. According to existing documents, the *solera* system was used in Spain after the Napoleonic Wars and began to spread abroad after the second half of the nineteenth century (Simpson 2003; Bonin 2002). It is therefore possible that this maturing system spread from Spain to Italy during the Cisalpine Republic, a French client republic founded by Napoleon Bonaparte in northern Italy, and which lasted from 1797 to 1802, mediated by the French oenological culture. Further historical research in this area would be very helpful to establish whether the *rincalzo* and *solera* systems are in some way related or developed independently in Spain and Italy (Giudici et al. 2009).

Only in the nineteenth century do a number of sources appear that refer directly to "balsamic vinegar", together with the first recipes and guidelines for making it using grape must alone. At the beginning of the century, vinegar making was a very popular activity widespread even among the lower classes. From this period onwards there is growing documentation of balsamic vinegars, increasingly detailed and reliable and representing the earliest testimony of the production methods that are still followed today (Saccani and Ferrari Amorotti 1999).

In 1806 Filippo Re (1763–1817), a botanist and agronomist from Reggio Emilia who had published books on several agricultural subjects that became very popular in his time, in a treatise titled *Elements of agriculture* (*Elementi d'agricoltura*), described a method for manufacturing a *Modena style vinegar* (*aceto alla modenese*), which is so similar to Mitterpacher's text that it appears clear he copied his predecessor:

> Modena vinegar is famous and popular. To manufacture it you boil white grape must in a vat for three days together with the stalks. The must is then transferred into a separate container, leaving the stalks apart in the vat, until they warm up, which will happen within three days. Afterwards, they are blended again with the grape must, leaving the mixture to lie for eight days. Then the must is transferred to separated tubs and a heated tile is put into each one. The must is transferred into casks when it has become vinegar, and after one year it is transferred again. But since the vinegar manufactured in this way is very strong, it is better to top it up often with an old vinegar, good in quality and mild. (Re 1806).

An interesting document appeared in the year 1819, an inventory of the vinegars in the royal storerooms (*sortita degli aceti dei Reali Magazzini*) of the Duke of Modena. The vinegar inventory was divided into *Balsamico* (Balsamic), *Mezzo Balsamico* (Half-balsamic), *Fino* (Fine), and *Ordinario* (Ordinary). It is not possible to be sure of the exact nature of these products, but it is an important indication that different qualities and types of vinegar definitely existed, possibly intended for different uses. There is no description of each vinegar type, but it is logical to deduce that *Balsamico* was the highest quality, that *Ordinario* was a low quality balsamic or possibly wine vinegar (without added grape must), and the other two grades were intermediate.

In 1839, the botanist and agronomist Giorgio Gallesio, who was interested in various fields, was passing sometime in the Modena area and visited the family *acetaia* at Count Salimbeni's farm (Rangone and Giacobazzi 2011). He wrote that

> the vinegar of Modena is of two sorts: vinegar from must and vinegar from wine, and it has yet to be decided which one is the best [...]. The only difference between grape must vinegar and wine vinegar depends on the first stage of production. I have seen that the must vinegar starts from cooked grape must, while the wine vinegar starts from wine, which is uncooked grape must. However, this is white wine and made from dedicated grape varieties, and it is made as clear as possible. Then, the same process is followed for both, keeping it in the sun for six months inside clay flagons and subsequently transferring it into casks [...]. (Gallesio and Baldini 1995)

During the course of the nineteenth century balsamic vinegar started to become an object of public trade, and with its high value many people clearly thought it offered good business opportunities. Salesmen and merchants wanted to sell it at markets, but because of the way balsamic vinegar was produced at the time, it was only available in small quantities. Consequently, some entrepreneurs started to look for a cheaper substitute for the original very expensive traditional balsamic vinegar that could offer similar sensory features, but that could also be produced more easily for trading on a larger scale. This search for a "quick" balsamic vinegar resulted in the creation of what might be called *artificial* balsamic vinegars, which started to spread at first very slowly, but increasing as the market demand expanded. Some written recipes and production methods began to appear, with a manuscript dated 1843 including a recipe. The document is preserved by Professor Luigi Amorotti, a famous scholar of Modena's local history. According to some authors, this is the oldest recipe of its kind ever found (Benedetti 1991; Rangone and Giacobazzi 2011). The manuscript clearly indicates an "*artificial balsamic vinegar (aceto balsamico artificiale)*", as follows:

> **Artificial Balsamic Vinegar**
> Boil 8 pounds of *saba*[7] of perfect quality in a clay or well tinned copper pot, until the consistency becomes good enough [...]. Then add 1 scrupulum[8] of powdered liquorice, 1/4 of cloves, 1/8 of the finest cinnamon powder, then add 3 steins[9] of the strongest vinegar and blend it altogether. Then, take a handful of matches, ignite them and soak them while blending. If the result is too sweet in taste, add new vinegar, or if it is too strong, add more *saba*.

From this time onwards, numerous similar recipes began to circulate in the area of Modena and Reggio Emilia. Nearly all of them were anonymous, hand written, and undated, but each one attests the fact that a quicker and easier method to produce balsamic vinegar was required, mostly for business reasons (Rangone and Giacobazzi 2011).

---

[7] *Saba* is a customary name for cooked grape must.

[8] *Scrupulum* is an ancient measure of weight, 1/24 of a Roman pound (approximately 1.296 g).

[9] The stein, or beer stein, was a customary measurement unit. Probably it was similar to the Imperial Pint, that is 568 mL (retrieved from the Official text of the Units of Measurement Regulations 1995, as originally enacted or made within the United Kingdom: http://www.legislation.gov.uk/uksi/1995/1804/schedule/made).

Another example is a recipe entitled "Recipe to make artificial balsamic vinegar in 6 months at most (*Ricetta per fare aceto balsamico artefatto ed in sei mesi al più*)":

Recipe to make Balsamic Vinegar artificially and in six months at most

Put 2 steins of very strong local vinegar into a jar of larger capacity.
Keep eight ounces of fine and well purified[10] *saba* apart.
Half a ounce of juniper extract.
One ounce of Jamaica pepper,[11] just crushed.
Four ounces of liquorice juice.
The skin of two fresh lemons.
A bunch of scented herbs like sage, spearmint and other aromatic species.

Put everything into a pot, heat on a low fire until the mixture boils gently, keeping it simmering for a while. Afterwards strain through a canvas. When it has cooled, add vinegar and blend. Then put the pot in the sun, well-sealed with a cork, turning it often so that it is warmed on all sides. It will be ready in a few months.

The annals report that in 1859 King Vittorio Emanuele II of Italy and the Prime Minister Camillo Benso di Cavour visited the Duke's *acetaie* (vinegar attics) in Modena, because by the end of the nineteenth century it was already acknowledged that the vinegar manufactured in this city was of special quality, based on knowledge handed down from ancient times and worthy of preservation as a cultural treasure.

2.7 A New Balsamic Era (1860–Today)

In the time of the Duke his cellars produced vinegar, or more likely a range of vinegars, that differed significantly from the balsamic vinegars we are familiar with today. Many of them were based on uncooked grape musts blended with sweet wines, creating very high levels of acidity and requiring many years of maturation before they could be used (Rangone and Giacobazzi 2011). In 1860 (or according to other scholars, 1862) there was a ground breaking event for the entire balsamic vinegar world and its future development: Francesco Aggazzotti (1811–1890), lawyer, agronomist, and politician in Modena, for the first time wrote down a detailed procedure for obtaining balsamic vinegar using only cooked grape must. He described this unusual product in three letters. The first, not very long and full of corrections and marginal annotations, was addressed to Pio Fabriani, dated by some scholars to 1860 and by others to 1862 (Fig. 2.6).

[10] Well purified *saba* means cooked grape must (saba) well filtered, in order to remove solid residues (tartaric acid crystals, at most).

[11] Jamaica pepper is allspice, also called myrtle pepper, pimento, English pepper or newspice. It is the dried unripe berry of *Pimenta dioica*, a tree native to the Greater Antilles, southern Mexico, and Central America.

Fig. 2.6 A portrait of Francesco Aggazzotti (1811–1890)

He wrote a more detailed and accurate second letter to Pio Fabriani on the 3rd of March 1862. The third letter was addressed to Ottavio Ottavi, enologist in the province of Alessandria, dated 1863. In these documents Agazzotti describes the process for making Balsamic Vinegar he was familiar with, and this is used as the basis for the current production regulations for the two Traditional Balsamic Vinegars of Modena and Reggio Emilia, definitively set out in the PDO (Protected Denomination of Origin) regulation, as established in 2000 by the European Community with EC Council Regulation No 813/2000.

While Aggazzotti was recording his method for making this precious vinegar, other scholars and enthusiasts left their own testimonies. An example is Fausto Sestini (1839–1904), a chemist and agronomist born in the province of Florence, who travelled a lot for his work. When he encountered the vinegar produced in Modena and Reggio Emilia, he was curious enough to start studying the subject and wrote a dedicated treatise in 1863 entitled *Sopra gli aceti balsamici del modenese* (Concerning the balsamic vinegars of Modena). One passage is particularly interesting:

> In the provinces of Modena and Reggio Emilia since ancient times a particular kind of vinegar has been prepared, the physical appearance and excellent aroma of which has earned it the name of Balsamic Vinegar. Among the elements of value of this product, age prevails over all. The people of Modena take a great care in authenticating it correctly. Balsamic Vinegars of 50, 70 or 80 years are considered to be good; those of 100, 120 or 150 excellent; and when, as happens frequently, their age is 200 or more years, there is nothing else more desirable! (Sestini 1863)

Sestini ascribes immense periods of maturation to the traditional balsamic vinegars he encountered, of over a 100 and even 200 years. As is now known, such extensive maturation is actually impossible even for traditional balsamic vinegars. There is a limit to the maturation period after which the product solidifies completely (Lemmetti and Giudici 2010; Falcone 2010). Rather than the age of the vinegar inside the barrels, we can only assume that Sestini was referring to the total age of the barrel set (*batteria*).

The tradition of developing recipes to obtain a balsamic vinegar similar to a traditional one, but in a few years, appears to persist after the Aggazzotti method. In 2013 one of these recipes was found in a manuscript preserved in a private house in Modena. Several traditional culinary preparations are described: quince jam, soda, ice cream, a "Modena-style" wine, and an "Artificial Balsamic Vinegar to be obtained in a few years".

From the handwriting of the author, who still remains anonymous, and from other details in the text, like choice of listed ingredients, and units of measurement used (dramas and ounces), it can be tentatively estimated that the manuscript was written at the end of nineteenth century or the very beginning of the twentieth century. Like the other recipes mentioned above, this one is very interesting, even if very short. Firstly, the Balsamic Vinegar is classed as "artificial, in a few years": it proposes a blend of cooked grape must and wine vinegar, with the addition of nutmeg, cloves, cinnamon, and coriander. This recipe would probably produce a density of about 1.20–1.25, flavoured with spices, and it was still left to mature for a few years. Translation of the text (Fig. 2.7):

To make the artificial balsamic vinegar in a few years, according to Prof. Barani

Common vinegar, the most acidic quality, 4 steins
Evaporated grape must, nearly syrup, 4 steins
These must be blended, and then add:
Nutmeg: 2 dramas
Cloves: 1.5 dramas

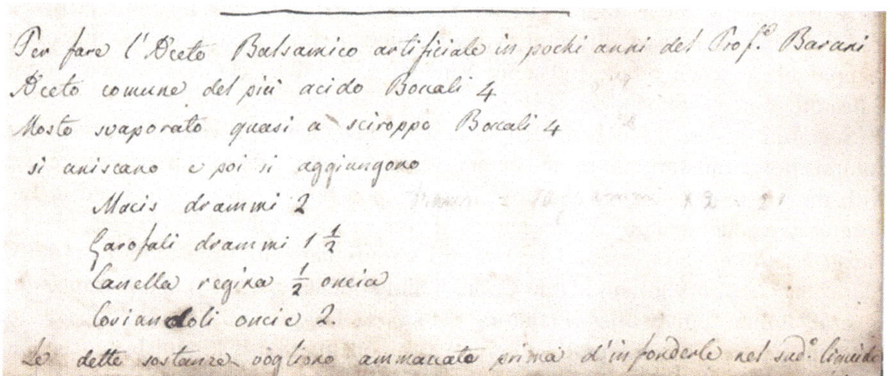
Per fare l'Aceto Balsamico artificiale in pochi anni del Prof.re Barani
Aceto comune del più acido Boccali 4
Mosto svaporato quasi a sciroppo Boccali 4
si uniscano e poi si aggiungono
Macis dramme 2
Garofali dramme 1 1/2
Cannella regina 1/2 oncia
Coriandoli oncie 2.
Le dette sostanze vogliono ammaccate prima d'infonderle nel sud.° liquido

Fig. 2.7 The manuscript found in a private house in Modena, describing a recipe for "artificial Balsamic Vinegar" (anonymous manuscript, private collection)

Cinnamon: 0.5 ounce
Coriander: 2 ounces
The above ingredients must be ground before infusing them in the liquid.
(Anonymous manuscript, private collection)

The discoveries discussed above demonstrate that for centuries Balsamic Vinegar was not defined by fixed procedures nor regulations of any kind. It was manufactured following several possible methods and using different raw materials. It appears that prolonged maturation and ageing of cooked grape must, fresh grape must, wine, *agresto*, apple juice, pear juice, mulberry juice, and other juices, if conducted correctly, always ends up producing a quality that might be termed "balsamic".

It also emerges that the culinary history of Modena and Reggio Emilia appears to have included two types of product that have always coexisted together through the course of history: a vinegar made with cooked grape must alone, which recently became the TBV we know today, produced on a small scale and very expensive, and secondly a vinegar made from grape must with added wine vinegar and/or with other ingredients (e.g. spices), this being cheaper and intended for a wider public and larger scale marketing, becoming the contemporary BVM and developing into a fully fledged industrial product. It should also be noted that the names of balsamic vinegars through the centuries have never been clearly defined in the minds of the general public, and the existence of these two "classes" of balsamic vinegars is proven at least since the eighteenth century (Saccani and Ferrari Amorotti 1999), more than 100 years before true industrialisation. To add further confusion, before the current distinction between TBVs and BVMs, balsamic vinegar of Modena was referred to in historical documents under names like *aceto alla modenese*(Modena-style vinegar), *aceto da padrone* (master's vinegar), *aceto del duca* (Duke's vinegar), *aceto balsamico naturale* (natural balsamic vinegar), and others.

The scenario today is much more transparent. First, there is the family tradition, maintained for centuries by thousands of households in Modena and Reggio Emilia, of producing vinegar in the attic, the main purpose not being commercial, but for daily use at home and as a gift to friends and relatives. Alongside this uninterrupted tradition, there are two different types of vinegar produced for sale, not even comparable to each other: Balsamic Vinegar of Modena (PGI) and Traditional Balsamic Vinegar of Modena (PDO).

Scientific research has provided significant insight into several aspects of the composition, transformation, and temporal evolution of these vinegars, together with their complex sensory features. Numerous scholars have contributed to the progressive understanding of these unusual vinegars. After the pioneering studies of Fausto Sestini in 1863, research on balsamics continued with Ernesto Parisi (1928), Mario Sacchetti (1936), and Dino Coppini and colleagues (1973) and continues in several Italian universities including Modena, Reggio Emilia, Bologna, and Pescara. Fausto Sestini identified certain components that distinguish the secular traditions of Balsamic vinegars from each other, not only for specific aromas but also regarding density, sugar content, and solid content. He identified the presence of substances like acetic acid, formic acid, acetate and formate ester, tartaric acid; intermediate forms of sugar dehydration; and humic acids, calcium and iron humate as responsible for the brown colour of balsamic vinegar.

Ernesto Parisi investigated the microbiology of the products, showing that the various transformations that occur in the formation of balsamic vinegar are due to yeasts and acetic bacteria. Mario Sacchetti established the biological nature of the transformation process of Traditional Balsamic Vinegar, succeeding in isolating *Osmophilus* yeasts and establishing the maximum concentration tolerance for sugar and alcohol content, highlighting how up to 50 % glucose can be developed. Starting in 1973, Professor Dino Coppini and his collaborators at the University of Modena performed numerous investigations on samples from century old barrels. Their research highlighted the presence of diacetyl, acetoin, 2,3-butylene glycol, and acetaldehyde (confirming the previous studies by Parisi). They also examined the presence of histidine and histamine, and investigated the volatile substances involved in aroma perception. Their studies mainly served to underline the major differences observed between traditional balsamic vinegar and any other type of vinegar. Research has been on going at the University of Modena and Reggio Emilia for over a decade (Saccani and Ferrari Amorotti 1999).

After two main kinds of balsamic vinegar had been defined by science, approaching the current era, their history consists largely of legal definitions, firstly issued in Italian regulations, and then more recently acknowledged by the European Commission. After an initial period when the Italian government legally protected the name "Aceto Balsamico di Modena", TBV was granted Protected Denomination of Origin (PDO) status by the European Community on April 17, 2000 with EC Council Regulation No 813/2000, in which the Production Regulations proposed by the Consortia were accepted, both for Modena and for Reggio Emilia TBV, and published on May 15, 2000.

Similarly, BVM was granted Protected Geographical Indication (PGI) status on 3 July 2009, with Commission Regulation (EC) No 583/2009, indicating composition, processing, maturation, packaging, and marketing of the more widespread *Aceto Balsamico di Modena*.

Notwithstanding the necessary legal and practical simplifications, the present chapter demonstrates that balsamic vinegars originated from a very long, time-honoured tradition and have evolved into their current familiar form only after an incredibly complex historical evolution.

References

Benedetti B (1991) L'aceto balsamico del Ducato di Modena. Il Fiorino, Modena

Bonin H (2002) Transnational companies, 19th-20th centuries. European Business History Association, P.L.A.G.E., Paris

Cato H (Tr.) (1591) Agricoltura nuova e casa di villa, di Carlo Stefano Francese. Aldo Manuzio, Venice

Columella LIM (I century CE) De re rustica. Public domain
Donizo (2008) Acta Comitissae Mathildis. Jaca Book
Estienne C, Liébault J (1570) L'agriculture et maison rustique. Chez Jaques du Puys, Paris
Falcone PM (2010) Crystallization and jamming i the traditional balsamic vinegar. Food Res Int 43:2217–2220
Gallesio G, Baldini E (eds) (1995) I giornali dei viaggi. Transcribed and annotated by Baldini E. Accademia dei Georgofili, Florence
Giudici P, Rinaldi G (2007) A theoretical model to predict the age of traditional balsamic vinegar. J Food Eng 82:121–122
Giudici P, Solieri L (2008) Vinegars of the world. Springer, Milan, pp 1–16
Giudici P, Gullo M, Solieri L, Falcone PM (2009) Technological and microbiological aspects of traditional balsamic vinegar and their influence on quality and sensorial properties. In: Taylor SL (ed) Advances in food and nutrition research, vol 58. Academic, San Diego, CA, p 137
Köbler G (2006) Neuenglisch-althochdeutsches Wörterbuch. Available at http://www.koeblergerhard.de/germanistischewoerterbuecher/althochdeutscheswoerterbuch/neuenglisch-ahd.pdf. Accessed 19 Jan 2014
Lemmetti F, Giudici P (2010) Bilancio di massa ed età dell'aceto balsamico tradizionale. Industrie delle Bevande 230:18–28
Mazza S, Muruoka Y (2008) Vinegar through the ages. In: Giudici P, Solieri L (eds) Vinegars of the world. Springer, Milan, pp 17–39
Mitterpacher L (1784) Elementi d'agricoltura. Imperial Monastero di S. Ambrogio Maggiore, Milan
Official text of the Units of Measurement Regulations (1995) Available at http://www.legislation.gov.uk/uksi/1995/1804/schedule/made. Accessed 19 Jan 2014
Palladius RTA (IV century CE) Opus agricolturae. Public domain
Plinius Secundus G [Pliny the Elder] (I century CE) Naturalis Historia. Public domain
Provvedimento 15 maggio 2000: Disciplinare di produzione della denominazione di origine protetta "Aceto balsamico tradizionale di Modena", published on Gazzetta Ufficiale della Repubblica Italiana n. 124, 30 maggio 2000
Provvedimento 15 maggio 2000: Disciplinare di produzione della denominazione di origine protetta "Aceto balsamico tradizionale di Reggio Emilia", published on Gazzetta Ufficiale della Repubblica Italiana n. 124, 30 maggio 2000
Rangone U, Giacobazzi G (2011) Dall'aquila estense al Tricolore. Tecnograf, Reggio Emilia
Re F (1806) Elementi d'agricoltura, 3rd edn. Stamperia Vitarelli, Venice
Regulation (EC) No 583/2009 of 3 July 2009 entering a name in the register of protected designations of origin and protected geographical indications [Aceto Balsamico di Modena (PGI)]
Regulation EC No 813/2000 supplementing the Annex to Commission Regulation (EC) No 1107/96 on the registration of geographical indications and designations of origin under the procedure laid down in Article 17 of Regulation (EEC) No 2081/92
Saccani F, Ferrari Amorotti V (1999) Il balsamico della tradizione secolare. Artestampa, Modena
Sestini F (1863) Sopra gli aceti balsamici del modenese. Bologna
Simpson J (2003) Spanish agriculture: the Long Siesta, 1765-1965. Cambridge University Press, Cambridge
The American Heritage® Dictionary of the English Language, 4 edn. Houghton Mifflin Company (2004). Available at http://dictionary.reference.com/browse/balsam and http://dictionary.reference.com/browse/balm. Accessed 15 Jan 2014
The American Heritage® Dictionary of the English Language, Appendix II: "Semitic roots", 4th edn (2000). Available at http://www.bartleby.com/61/roots/S54.html. Accessed 19 Jan 2014
The Merriam-Webster Online Dictionary. Available at http://www.merriam-webster.com/dictionary/balsam. Accessed 19 Jan 2014

Vocabolario degli Accademici della Crusca in rete. Available at http://vocabolario.biblio.cribecu.sns.it/Vocabolario/html/_s_index2.html. Accessed 15 Jan 2014

Zupko RE (1981) Italian weights and measures from the middle ages to the nineteenth century, vol 145. American Philosophical Society, Philadelphia, PA

Chapter 3
The Balsamic Family

3.1 Balsamic Family: Legal Definitions

As noted in the previous chapter, the term "balsamic", from an etymological perspective, has a clear and unambiguous meaning in all the modern languages of Europe and the Americas, deriving from its shared origin in ancient Latin and Greek and the common Semitic root of the word. In the terminology of foodstuffs, the healing connotation of the term *balsamic* is generally outdated and the sense is restricted to the sensorial aspects of balsamic aromas that a substance is thought to emanate. In the specific case of vinegars, a precise, unambiguous definition of what the balsamic attribute means, and what exactly balsamic vinegars are, is still far from being universally agreed. Giudici et al. (2009) described the technological and microbiological aspects of Traditional Balsamic Vinegar and their role in the quality and sensorial properties.

Generally speaking, there is a wide range of vinegars, all around the world, made from grape must and with the generic and legally unsupported definition of "balsamic vinegars". The market for balsamic vinegars and related products has developed very rapidly over a short interval of time, today comprising a wide variety of products that, at first glance, might appear quite similar both in appearance and from a sensory perspective. However, all these products can vary enormously as regards ingredients, market claims, price, and legal status. Understanding the real differences between the members of this heterogeneous group of products is not easy. Consequently, a summary of what might be called the *Balsamic Family* is set out below. An initial clarification regards the legal definition of each group of products and three consequent levels of legal protection are identified.

Condiments This group is composed of products that cannot be defined as vinegars because of their composition, low acidity, or intended use. They have no compositional limitations and can contain thickeners, preservatives, colours, flavours, and any kind of additive, both artificial and natural. They can be liquid, solid,

P. Giudici et al., *Balsamic Vinegars*, DOI 10.1007/978-3-319-13758-2_3

or semisolid. Sometimes they resemble vinegars and even balsamic vinegars in many aspects. Their pricing and qualitative standards vary widely and depend on the cost of the initial raw materials. Balsamic sauces, glazes, jellies, flavoured vinegars, oil and vinegar emulsions, fruit and vinegar mixes, vinaigrettes, and similar products fall into this group. These products are disciplined by general alimentary laws, according to national and international regulations.

Vinegars These are a subcategory of the larger condiments group. According to a generally accepted definition, vinegars are liquids obtained from the acetic fermentation of any suitable foodstuffs, with or without previous alcoholic fermentation. Vinegars are legally defined in many countries, often with specific legislation, usually imposing minimum acidity levels. For example, in Europe vinegars must contain from a minimum of 5 g to a maximum of 12 g of acetic acid per 100 ml of product, with the exception of wine vinegar, which must contain a minimum of 6 g per 100 ml. In the United States, the minimum is 4 g per 100 ml, and in Australia 4 g per 100 g. Vinegars are usually inexpensive, everyday condiments, but their prices can be higher if they require a complex production process and extended maturing or are obtained from high-cost raw materials (e.g. special wines). This group includes wine vinegar, rice vinegar, apple cider vinegar, malt vinegar, honey vinegar, etc., which are normally obtained from a single foodstuff through alcoholic and subsequent acetic fermentation. In some countries, flavoured vinegars are included in this category.

Speciality Condiments and Vinegars This group includes vinegars and condiments that are legally recognised as distinctive and different from the products of the "normal" vinegars or condiments group, for historical, cultural, or other significant reasons. These products are subject to specific regulations or GI, for example PGI or PDO in Europe. The vinegars and condiments in this group are often, but not always, expensive and are normally produced on a limited or small scale, like the TBV. However, there are cases of very large production volumes, with examples including Jerez Vinegar and Balsamic Vinegar of Modena (not to be confused with TBV), which today are industrial products consumed worldwide, like wine vinegar.

Products of the balsamic family can belong to any of the groups above, as shown in Fig. 3.1, which illustrates the different levels of legal regulation of condiments and vinegars, as well as the distribution of the various products in the balsamic family.

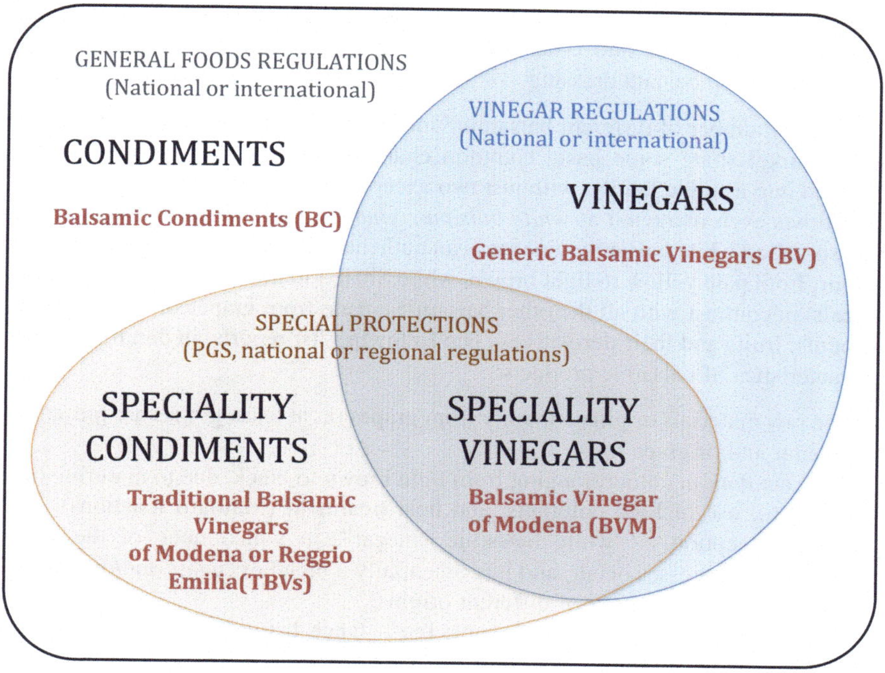

Fig. 3.1 Different levels of legal status within the vinegar market. Balsamic family products (in *red*) belong to any of the defined groups

3.2 Composition and Main Features of Balsamic Vinegars and Condiments

What is intended nowadays as *balsamic* vinegars and/or condiments are dark glossy products, with a distinctive sweet and sour taste. Their density and viscosity can vary depending on commercial quality, but, generally, consumers associate the *balsamic* status with thickness. The sweetness from sugars modulates the acid taste, mostly due to acetic acid, and reduces the perception of sourness.

In order to establish a set of essential characteristics shared by the balsamic family, relevant data was collected from a wide range of sources. These sources included:

- Direct knowledge, resulting from years of study of industrial and traditional balsamic vinegars.
- A collection of balsamic vinegars purchased in Italy, Germany, France, Spain, Japan, and the United States.

- Web searches using the word balsamic in various languages, associated with the names of countries and other keywords, such as vinegar, sauce, seasoning, condiment, spice, and dressing.

A large number of different balsamic vinegars and condiments available on the retail market share some basic common characteristics. Therefore, they can be grouped into a single family, with just two exceptions: a white condiment, which is sometimes even marketed as *white balsamic vinegar*, and Shanxi mature vinegar. The so-called white balsamic vinegar is an authentic oxymoron because of its light colour, from pale yellow to light brown, while Shanxi mature vinegar is made from cereals, in contrast with all the other balsamics made from grapes or, in very small amounts, fruits and their derivatives. The following list reports, in detail, the main characteristics of balsamic products:

- The raw materials originate mainly from grapes or, at least, grapes are present as vinegar and/or grape must.
- They are dark in colour, ranging from light brown to black, due to maturing (see Chap. 6), use of food colorants, and heat treatment (Maillard reactions). The single exception is "white balsamic vinegar", in which none of the listed darkening processes occur, and it is essentially a blend of sugars and/or concentrated juices and vinegar of different origins.
- They have a sweet and sour flavour: the balance between these two primary tastes is the most important sensory attribute in consumer preferences.
- They give a mild perception of acidity: although the average concentration of acetic acid is similar in most products on the market, the general aim is to mask it, because an excess of sourness is a negative trait in most cultures, or it is associated with low-class (cheap) balsamic vinegars.
- Viscosity is variable and increases with the cost of the product. Vinegars and balsamic vinegars for everyday use are relatively runny, while high-end balsamics are syrupy and thicker. Condiments designed to imitate or extend the balsamic vinegar range exhibit different viscosities, depending on their commercial target.
- Density ranges from 1.060, which is the minimum level for a legal PGI *Aceto Balsamico di Modena*, to a maximum around 1.38. Theoretically, there is no upper limit for density, but the oldest Traditional Balsamic Vinegar (TBV), after extended maturing, typically shows physical defects when density reaches the 1.37 level, such as precipitation of tartrates and sugars, loss of viscosity and even acidity, and complete solidification in the worst cases (Falcone and Giudici 2010; Falcone 2010).
- Intended use is the most subjective aspect and therefore the most variable, depending in part on product features but also largely on personal habits, culture, and the immediate needs of the consumer. Uses range from day-to-day salad dressings as a simple substitute for common table vinegar, to enhance the flavour of seasoned cheese, or inclusion in traditional cooking recipes.

The authors collected and carefully examined a large volume of analytical data in order to set out an evaluation protocol that takes into account: national and

Fig. 3.2 A small group of balsamic products from our collection

international legislation, intended use, ingredients and additives, and production technology. Based on these variables, four categories of balsamic products were identified:

- Generic balsamic vinegar, with no reference to any recognised geographic origin, simply called Balsamic Vinegar (BV).
- Balsamic Vinegar with Protected Geographical Indication (PGI), Balsamic Vinegar of Modena (BVM).
- Balsamic Condiment with a higher level of legal protection, the Protected Designation of Origin (PDO), Traditional Balsamic Vinegar of Modena or Reggio Emilia (TBV).
- Balsamic glazes, sauces, jellies, and other food dressings without any legal protection, most appropriately called Balsamic Condiments (BC).

Figure 3.1 illustrates the different levels of legal regulation of condiments and vinegars, as well as the distribution of the various products in the balsamic family. Figure 3.2 shows a small collection of balsamic vinegars from all over the world.

3.2.1 Generic Balsamic Vinegars

The category of generic balsamic vinegars (BV) includes products that must comply with general vinegar regulations, which vary from country to country, without any other constraints concerning composition and maturing. BV includes flavoured vinegars for everyday use, with low or high viscosity and density, but always characterised by a prominent acidity that identifies them as *authentic* vinegars. The BV resemble BVM for their intended use and sensorial properties, but they may differ in certain components and additives not permitted in the production of BVM, with its strictly regulated production process under the GI requirements.

3.2.2 Balsamic Vinegar of Modena

Balsamic Vinegar of Modena (BVM) is the English translation of the protected name *Aceto Balsamico di Modena*, a product manufactured according to a standard of identity submitted to the European Commission in order to obtain recognition of Protected Geographic Indication (PGI), awarded in July 2009. In summary, BVM is formulated by mixing cooked and/or concentrated grape must, at least 10 % of wine vinegar, a small amount of vinegar aged 10 years, and caramel colouring up to 2 %. The percentage of cooked and/or concentrated grape must is required to be more than 20 % of the total volume. No other substances are allowed. The resulting blend is refined in wooden containers (barrels, casks, or vats) for at least 60 days in order to be certified as PGI Aceto Balsamico di Modena. The production process, from blending of raw materials to refining and maturing in wooden barrels, must take place within the geographical area of origin, inside the provinces of Modena and Reggio Emilia. Once the product has been PGI certified, packaging can occur outside the area of origin. The packaging of the final product must bear the name *Aceto Balsamico di Modena* along with the phrase *Indicazione Geografica Protetta* (protected geographical indication) written in full or as an acronym. No other qualifications are allowed, even in numerical form. Only the word *invecchiato* (matured) can be added without any further additions, provided that the product has been matured for a period of 3 years or more in casks, barrels, or other wooden containers.

The chemical and physical features of BVM, as stipulated in Regulation CE 583/2009, are summarised in Table 3.1.

The sensory properties are also defined in the regulation, as set out in Table 3.2.

Table 3.1 Legal parameters of Balsamic Vinegar of Modena (BVM), according to Regulation CE 583/2009

Parameter	Value
Total acidity	>6.00 g/100 ml
Relative density 20°/20°	>1.060
Total alcohol	<1.5 ml/100 ml
Total sulphur dioxide	<100 mg/l
Ashes	>2.5 g/l
Net dried extract (excluding sugar)	>30 g/l
Reducing sugars	>110 g/l

Table 3.2 Sensory properties of Balsamic Vinegar of Modena (BVM), according to Regulation CE 583/2009

Parameter	Description
Clarity	Clear and bright
Colour	Deep brown
Aroma	Persistent, delicate, and slightly acidic with woody overtones - taste: bittersweet, balanced

Table 3.3 Legal parameters of Traditional Balsamic Vinegar of Modena and Reggio Emilia, according to Regulation CE 813/2000

Parameter	TBV Modena	TBV Reggio Emilia
Total acidity	>4.5 g/100 g	>5 g/100 ml
Relative density $d_{20/20}$	>1.240	>1.200

3.2.3 *Traditional Balsamic Vinegars*

Despite their name, Traditional Balsamic Vinegars of Modena or Reggio Emilia (TBV) are two PDO balsamic condiments obtained by a traditional method as described in Chaps. 4 and 5. The only ingredient is cooked grape must, fermented, acetified, and subsequently transferred in to a series of barrels and matured for at least 12 years, before bottling. A panel of trained tasters assesses the sensory characteristics of each batch of TBV. See Chap. 7 for further details.

The chemical and physical features of TBV, as stipulated in Regulations CE 813/2000 and 510/2006, are summarised in Table 3.3, while Table 3.4 shows the sensory properties.

3.2.4 *Balsamic Condiments*

Balsamic Condiments (BC) are the most heterogeneous group within the balsamic family. They include condiments that cannot be defined as vinegars, as noted above in Sect. 3.1.

The motivations for creating condiments are diverse. Producers may want to offer variants on existing products with different properties, e.g. lower acidity, different colours, flavours, higher viscosity, etc. Moreover, they might sell high-

Table 3.4 Sensory properties of Traditional Balsamic Vinegar of Modena and Reggio Emilia, according to Regulation CE 813/2000

Parameter	TBV Modena	TBV Reggio Emilia
Colour	Bright dark brown	Dark brown, clear, gleaming
Consistency/density	Like a runny syrup	Obviously runny and syrupy
Aroma	Characteristic, fragrant, and, altogether, well-formed, penetrating, and persistent bouquet with a noticeable but pleasant and harmonious acidity	Penetrating and persistent, fragrant, with pleasing acidity or characteristic bouquet depending on the wood used - oak, chestnut, cherry, juniper, mulberry, ash, and acacia - and long maturation periods
Flavour	Characteristic of balsamic vinegar, in line with an unchanging centuries-old tradition; sweet and sour and well balanced, with appreciable acidity and a hint of aroma imparted by the wood used for the vats; strong, clear, full, velvety, intense, and persistent, in keeping with its characteristic bouquet	Well-combined sweetness and sourness, distinct acidity, and aromatic in compliance with its olfactory properties

quality products at a reduced price, for example, by avoiding the maturing period. They may also experiment with new vinegar or balsamic-like products containing different raw materials, e.g. fruits, vegetables, honey, etc.

BC may resemble BVM or TBV for their commercial target. They can be fluid or very viscous, and they can be thickened as required using starch or other additives. BC can include mixtures of vinegars of different origins (spirit, wine, fruit, malt, etc.) together with grape must, fruit juices, sugars, and food additives to achieve the appearance and texture required for dressings and similar usage. These condiments are ready to use and often commercialised in special bottles that simplify their application, which makes them very popular among chefs and consumers.

Table 3.5 provides a comprehensive list of balsamic products grouped into the most important types of vinegar and vinegar-like condiments, categorised on the basis of the principal legal and technological aspects.

It is very difficult to define the precise differences within this heterogeneous group of products that includes condiments, vinegars, and speciality vinegars legally recognised and protected by specific regulations like PGI and PDO (Table 3.5). The majority of balsamic products are simple blends of two or more ingredients in different amounts and ratios, including vinegar, concentrated must, sugars, food colouring, and thickeners. TBV are the only ones produced by fermentation, more precisely by acetification and ageing of a single ingredient: cooked grape must. Furthermore, a legal definition for the three types of protected balsamic vinegars (BVM and TBV) only exists within the European Community, as discussed in Chap. 8. Outside the EC boundaries, there are no such regulations or monitoring bodies offering a similar level of control. Consequently, products labelled "balsamic" can differ widely from each other around the world.

Table 3.5 Classification of balsamic products on the basis of raw materials and ingredients, maturing, market legislation, legal acidity, and physical state

Product type	Added vinegar	Grape must	Added sugars	Caramel (E150a–d)	Thickeners
Vinegars					
Vinegars (apple, honey, malt, whey, etc.)		No	No	No	No
Wine vinegar		No	No	No	No
Balsamic Vinegar (BV)	*Yes*	*Yes*	*No*	*Yes*	*No*
Speciality vinegars					
Vinagre de Jerez (Spain)	Yes	0–50 % (also used as colour)	No	No	No
Vinagre del Contado de Huelva	Yes	0–50 % (also used (as colour)	No	No	No
Aceto Balsamico di Modena (BVM)	*Yes >10 %*	*>20 %*	*No*	*Yes 2 % max*	*No*
Condiments					
Liquid condiments not classifiable as "vinegar" or "balsamic"	Depends on recipe	Depends on recipe	Permitted	Depends on recipes	Permitted (usually not used)
Balsamic condiments (sauces, glazes, or jellies) (BC)	*Permitted depends on recipe*	*Depends on recipe*	*Modified or native starch (maize, potato, wheat, etc.) Glucose/fructose syrup (fluid or dehydrated) Pectins, dextrose, gums (xanthan, guar, etc.), carob seed flour, lactose*	*Permitted*	*No*
Aceto Balsamico Tradizionale di Modena (TBVM)	*No*	*100 %*	*No*	*No*	*No*
Aceto Balsamico Tradizionale di Reggio Emilia (TBVRE)	*No*	*100 %*	*No*	*No*	*No*

Product type	Flavours	Maturing	Market protection	Legal acidity limit	State
Vinegars					
Vinegars (apple, honey, malt, whey, etc.)	Natural flavours only	Not mandatory	No	EU 5–12 % (w/v) USA > 4 % (w/v) AUS > 4 % (w/w)	Liquid

(continued)

Table 3.5 (continued)

Product type	Flavours	Maturing	Market protection	Legal acidity limit	State
Wine vinegar	Natural flavours only	Not mandatory	No	EU 6–12 % (w/v) USA > 4 % (w/v) AUS > 4 % (w/w)	Liquid
Balsamic Vinegar (BV)	*variable*	*Not mandatory*	*No*	*not mandatory*	*Light to high bodied liquid*
Speciality vinegars					
Vinagre de Jerez (Spain)	No	Yes	PDO	>7 %	Light to high bodied liquid
Vinagre del Contado de Huelva	No	Yes	PDO	>7 %	Light to high bodied liquid
Aceto Balsamico di Modena (BVM)	*No*	*>60 days or >3 years for "invecchiato" claim*	*PGI*	*>6 g/100 ml*	*Light to high bodied liquid*
Condiments					
Liquid condiments not classifiable as "vinegar" or "balsamic"	Permitted	Not required (usually no)	No	No	Liquid
Balsamic condiments (sauces, glazes or jellies) (BC)	*Permitted*	*No*	*No*	*No*	*Thick liquid—Semisolid*
Aceto Balsamico Tradizionale di Modena (TBVM)	*No*	*>12 years*	*PDO*	*>4.5 g/100 g*	*High bodied liquid*
Aceto Balsamico Tradizionale di Reggio Emilia (TBVRE)	*No*	*>12 years*	*PDO*	*>5 g/100 ml*	*High bodied liquid*

Balsamic products are in cursive

3.3 Specific Traits of the Balsamic Family

3.3.1 The Colour of Balsamics

The colour is variable and ranges from amber brown to dark brown to nearly black. The colour tone and saturation can be the result of different processes:

- Maturing
- Addition of food colouring
- Heat treatment of grape juice, concentrated grape juice, or sugars and starch derivatives

The technology utilised to achieve the desired colour has a strong influence on the final cost of the vinegar. Extended maturing is very expensive and time-consuming and is restricted to the most expensive BVM and to TBV. Food colouring or heat treated starch derivatives are commonly used for BC and BV.

The colour of TBV is due to the presence of chromophore-labelled melanoidins that originate during the cooking of the must, subsequently increasing in concentration during maturing. At least four average sizes of melanoidins contribute differently to the colour. The Brown Index (BI), formulated as the ratio between the absorbance at 420 nm of TBV and the absorbance of the cooked and acidified grape must, has been proposed as an indicator of the increasing brown colour during maturing (Falcone and Giudici 2008). See Sect. 3.4.2 for further details. The most widespread colorant is caramel, specifically the sulphite ammonia type, also called type 4 and coded E150d in Europe. Caramel is produced by heating carbohydrates in the presence of ammonia salts and sulphites (possibly with ammonium sulphite). Caramel is a widely used ingredient in sodas and other beverages. In some countries, its use is limited in food production because of the presence of 4-Methylimidazole (4-MEI) and other nitrogen thermal treatment by-products, all substances recognised as potential carcinogens. These health concerns also regard vinegars and consequently some producers do not add caramel on their BC and BV, instead using alternative colours or technologies to achieve the dark-brown shade.

The third method to darken vinegar is heat treatment of the grape must or of the finished products, when dextrin and maltodextrin are usually added to increase the browning effect (see Chap. 4). A negative consequence of extreme thermal treatment is the formation of furfurals, the concentration of which is directly related to water activity, temperature, exposure time, and the pH of the treated product (Muratore et al. 2006).

3.3.2 Sweet and Sour Taste

The acid taste is strongly mitigated by sugars and many other compounds present in balsamic products, for example tannins, starch, melanoidins, and caramel.

The total acidity from acid–base titration is the sum of the contribution of volatile and non-volatile acids. The acids in balsamics are of different origins; in general the tartaric and malic acids derive from the raw materials, D- or L-lactic acid from lactic fermentation, succinic acid from yeast fermentation, and acetic acid from acetic acid bacteria (AAB). In addition, citric, DL-lactic, and DL-malic acids can be additives. It is worth noting that knowledge of the quantity of sugars and titratable acidity is not enough to predict the sensorial perception of sweet and sour flavour of balsamic products (see Chap. 7).

3.3.3 Pungency

Pungency is a characteristic trait of the flavour of vinegar and represents a simultaneous sensation of irritation, pain, and burning, which are some of the trigeminal stimulations felt during olfactory testing. In balsamic vinegar, pungency is due to the acetic acid in gaseous state in the headspace of the bottle. It is closely related to the volatile acidity of the vinegar and the temperature of samples during sensory testing. BC and TBV are usually less pungent than generic BV because of the lower concentration of acetic acid. Furthermore, balsamic consumers perceive high levels of pungency as a negative trait, because they associate high pungency with immature vinegars. Since BC are not matured, their pungency derives from the formulation of the condiment. Conversely, TBV maturing is a mandatory part of the production process and a considerable amount of acetic acid is lost through evaporation, especially if the maturing barrels are left open to improve the concentration process. In response to consumer preferences, and as the result of different production methods (see Chaps. 4 and 5), TBV's produced today are less pungent and sweeter than in the past.

3.3.4 Viscosity, Syrupiness, and Consistency

In the strictest sense, viscosity is a measurement of the strength of resistance of fluid molecules to shear forces, and it can easily be measured instrumentally. However, from the sensory point of view, viscosity relates to the tactile stimulation of the sensory system connected to the trigeminal nerve, which has both mechanical and chemical receptors. The terms syrupiness, viscosity, and consistency are often synonyms, with consistency referring to the syrupy effect due to viscosity and relative density. The presence of substances with high-molecular weight also contributes significantly to the rheological behaviour (see Chap. 7). For BC and TBV, viscosity, syrupiness, and consistency are important attributes, and the higher the viscosity, the higher the perceived quality by consumers. These sensorial properties are achieved in different ways: increasing relative density by adding sugars or high-density syrups, or the addition of thickeners such as starch. For TBV, the addition of any additives or raw materials other than grape juice is not permitted, and the most widespread practice is the use of high-density cooked must at the beginning of the process. High-density must is a shortcut to achieve TBV with high specific gravity, but not necessarily high viscosity. The latter practice also has the drawback of giving rise to vinegars with problems of physical stability (Falcone 2010; Falcone et al. 2011).

3.3.5 *Relative Density*

The relative density (RD), in physical terms, is the relationship between the density of a substance and the density of water at the same temperature, and it is also called specific gravity. In balsamic vinegars, RD depends on the nature and percentage of substances other than water. RD should not be confused with the viscosity as they are two different measurements: density is a measurement of the quantity of substance per volume unit, while viscosity is a measurement of the cohesive force within a fluid. The RD of vinegars is mainly due to the sugar content, which, in some BC, can be close to the solubility limit. RD of balsamic products is modulated by the addition of sugars and concentrated syrups and juices. Other practices are more sophisticated and involve the same techniques used in juice and food concentration, such as cryoconcentration and reverse osmosis. For high-quality TBV, the RD increases during maturing in a series of barrels: the wood of the barrels acts as a semipermeable membrane for the small sized molecules including water, while being impermeable to molecules of larger dimensions (see Chap. 4).

3.4 TBV Specific Properties

TBV's are vinegars characterised by their strong local identity as well as certain chemical–physical and sensory properties, which are defined by Italian legislation and recognised within European geographical indications as PDO. Local associations of producers and experts in the field, usually called consortia, protect and control TBV and its reputation. The aim of the existing consortia is to promote the culture of TBV and survey its production and distribution. There are currently three active Consortia in both the cities involved, Modena and Reggio Emilia, with the same purpose and functions. The Modena TBV consortia are the *Consorzio Produttori Aceto Balsamico Tradizionale di Modena*, *Consorzio Tutela Aceto Balsamico Tradizionale di Modena*, and *Consorteria dell'Aceto Balsamico Tradizionale di Modena*. The Reggio Emilia consortia are the *Consorzio fra Produttori di Aceto Balsamico Tradizionale di Reggio Emilia*, *Confraternita dell'Aceto Balsamico Tradizionale di Reggio Emilia*, and *Sindacato Produttori Aceto Balsamico Tradizionale di Reggio Emilia*.

TBV's are generally described as vinegars of ancient origin, possibly dating back to the Middle Ages and so with a long history, deeply embedded in the gastronomic traditions of the Italian provinces of Modena and Reggio Emilia. TBV's have achieved worldwide fame in recent years, highlighted in several marketing initiatives. Notwithstanding the claims of extremely long-standing traditions, as suggested by the names, the culture, history, and complexity underlying these products are far from being accurately reconstructed and fully understood (see Chap. 2).

The chemical composition of balsamic products only depends on the characteristics of the raw materials and their relative quantities: the amounts of sugar, concentrated grape juice, vinegar, thickeners, flavouring agents, and colouring used in the initial formulation. The blend can be further concentrated up to the desired density and viscosity (see Chap. 4). The sole exception is TBV, since this condiment is based on a single permitted raw material, cooked must, and the final composition is due to microbiological transformation and maturing. One of the main features of TBV is the maturing period, fixed at a minimum of 12 years for the youngest quality and a minimum of 25 years for the most refined and valuable quality. Many producers claim that their products are aged up to 30 years or more. Another factor that differentiates TBV from BC and BV is the homogeneity of composition within the group. The overall composition of TBV samples with regard to sugar concentration, titratable acidity, volatile acidity, and density all fall within a narrow range, while in BC there is much more variability. For this reason, only the composition of TBV is reported below.

3.4.1 Chemical Composition of TBV

Sugars

Glucose and fructose are the main components of TBV, and their relative quantity is close to 50 % (w/w). In the past, it was common practice to estimate the sugar content of TBV with a refractive measurement (Brix degree), but in reality, the two measurements are not related in any way, given that all vinegar components have a refractive response and solute-specific refractive coefficient. The fact that the refractive response of old TBV samples is higher than that of cooked must is likely due to the presence of high-molecular-weight biopolymers (Falcone and Giudici 2008). The reducing sugars, glucose and fructose, are present in isoforms, with only α and β pyranosidic forms present for glucose, but both pyranosidic and furanosidic α and β forms for fructose (Consonni et al. 2008). Other minor sugars have been detected in TBV. Expressed in g/kg they were identified as follows: xylose 0.11–0.39; ribose 0.078–0.429; rhamnose 0.061–0.195; galactose 0.136–0.388; mannose 0.41–1.46; arabinose 0.33–1.00; and sucrose 0.46–6.84 (Cocchi et al. 2006).

Organic Acids

In oenology, titratable acidity is conventionally expressed in grams of tartaric acid per litre of must or wine, while for vinegars it is expressed in grams of acetic acid per 100 ml of product, often indicated as percentage by volume. For TBV and BC with a specific gravity significantly greater than 1, it is more appropriate to express acidity by weight (g of acetic acid per 100 g of product). In grape must, titratable acidity coincides with fixed acidity due to the presence of tartaric and malic acids. The titratable acidity of the musts and the ratio of the two acids depend on the variety and ripeness of the grapes. Generally, the level of tartaric acid is constant during ripening and acidity decreases during ripening due to the consumption of malic acid. The level of tartaric acid in concentrated musts is further influenced by the low solubility of cream of tartar, the potassium salt of its conjugated base. Grape pressing and the cooking of must cause abundant precipitation of calcium and potassium tartrate salts, and this is accentuated during fermentation when the development of alcohol further reduces the solubility of these salts (see Chap. 5).

The organic acids in vinegar are, in descending quantitative order, acetic, tartaric, gluconic, L-malic, succinic, and D- and L-lactic and citric acids. AAB mainly produce acetic and gluconic acids through metabolisation of ethanol and glucose, respectively. Tartaric and malic acids originate from the grapes, succinic acid from yeast metabolism, and lactic acid, generally present in low concentrations, arises mainly from lactic acid bacteria but also from the thermal treatment of grape must. Malic, succinic, citric, and perhaps other organic acids can be oxidised in the presence of high levels of oxygen. Gluconic acid, for example, can be produced by AAB in the presence of glucose and after ethanol depletion, which is why it was proposed as an indicator for TBV authenticity (Giudici 1993). However, it can also be a constituent of grapes deteriorated by moulds, i.e. *Botrytis cinerea*.

Compared to wine vinegars, TBV samples have very similar titratable acidity levels, but a different qualitative composition. Tartaric acid is present in low percentages in mature TBV and its concentration decreases noticeably during maturing because of the precipitation in the form of potassium and calcium salts. Succinic and citric acids occur in relatively low concentrations and they reach asymptotic concentration levels when the TBV maturing process begins (Cocchi et al. 2002). Giudici (1993) demonstrated that the level of succinic acid tends to decrease slowly because this occurs in certain esterification reactions during the maturing process and that gluconic acid is present in a wide range of concentrations depending on the technological production method.

Volatile Compounds

The profile of volatile and aromatic fraction of TBV varies widely in relation to the measurement methods. This makes it difficult to compare data from different authors, laboratories, and procedures. In a quantitative and qualitative study of the volatile compounds in several kinds of vinegar, BV and TBV were found to contain various compounds including aldehydes, ketones, alcohols, and esters (Zeppa et al. 2002; Natera et al. 2003) A quantitative measurement of volatile compounds was used to group and discriminate vinegars of different origins (Del Signore 2001). Among the alcohol varieties, ethanol and 2-propanol were found in relatively high concentrations in common vinegars, while 1-propanol, isobutyl alcohol, isoamyl alcohol, and 1-hexanol were more common in BV. Acetaldehyde was found in relatively high concentrations in common vinegars, while diacetyl, hexanal, and heptanal were found in three and five times higher concentrations in TBV compared to BV and common vinegars. Propionic acid was found in BV and at relatively high concentrations in TBV. Esters were found in BV and common vinegars in higher concentrations than in TBV, apart from 1,3-butanediol diacetate.

Phenolic Compounds
The phenol contents of TBV samples were analysed both with colorimetric method (Verzelloni et al. 2007) and by gas chromatography-mass spectrometry (GC-MS) (Plessi et al. 2006), but the quantitative characterisation of each individual phenolic acid is still unsatisfactory. Natera et al. (2003) grouped and classified vinegars on the basis of phenolic composition, aroma compounds, and organic acids. A total of 83 vinegars of different origins and raw materials were analysed. Linear discriminant analysis allowed an 88 % differentiation based on raw material and 100 % on the presence or absence of maturing in wood. The antioxidant properties of traditional balsamic vinegars relative to phenolic and flavonoid content, in comparison with selected vinegars and red wines, were investigated by Verzelloni et al. (2007). The polyphenolic content was determined utilising Folin–Ciocalteu in peroxidase assay. The antioxidant capacity was quantified using both ABTS and FRAPS assays. The results show that traditional balsamic vinegar has higher antioxidant activity as well as phenolic and flavonoid content than all the other tested products, with the sole exception of Nero d'Avola red wine. The antioxidant capacity of wines and vinegars is closely correlated to their phenolic and flavonoid content, while in TBV and BV this correlation is lower.

Furanic Compounds
The 5-hydroxymethylfurfural (HMF) content of different vinegar samples, made from cooked must or other raw materials, was determined by Theobald et al. (1998). The authors concluded that the vinegars studied could be divided into four groups depending on their HMF content: samples with zero, low, medium, and high HMF concentrations. The balsamic vinegars had a concentration ranging from 300 mg/l to 3.3 g/l of HMF, with a detection limit of 80 μg/l. The highest concentrations were found in TBV samples, and up to 5.5 g/kg for the samples claimed to be matured for extended periods. A quantitative analysis of furanic compounds in different Italian vinegars was carried out by Giordano et al. (2003). The authors developed a method based on headspace solid-phase micro-extraction coupled to GC-MS using isotope dilution with a detection limit of 15 mg/l after the effects of ionic strength, temperature, and extraction time had been optimised. The authors found 2-furfural ranging from 14.2 to 2.6 mg/l and 5-methylfurfural from 2.4 to 1.1 mg/l. A comparative study, focused on the accumulation of furanic compounds including HMF in various commercial BV brands from Modena province, was recently presented by Bononi and Tateo (2009). The authors revealed that the furan levels in the majority of the 21 samples purchased on the Italian market ranged between 4 and 26 ng/g.

The term *"traditional"*, referred to a food product, cannot be used to justify freeing producers or authorities from the responsibility to assess for potential safety risks related to its production, including the HMF content. However, it is reasonable to suppose that, since TBV is generally used as a condiment, its effective contribution to daily HMF intake should be carefully evaluated before it can be considered a serious risk to public health (O'Brian and Morrisey 1989; Janzowski et al. 2000; EFSA 2014).

Melanoidins and Other Biopolymers
The chemical properties and other characteristics of melanoidins are believed to derive from their complex molecular structure and size, which have not yet been defined because they can change considerably depending on the source of reactants and the reaction conditions. The first attempt to characterise TBV production as a function of the molecular size distribution of its constituents was published recently by Falcone and Giudici (2008). The authors suggested that TBV can be treated as a heterogeneous mixture of small molecules and copolymers with molecular weights from 0.2 kDa to over 2,000 kDa, highly polydispersed with regard to their molecular size and composition of chromophore label, ranging from uncoloured to brown structures. The relative concentration of melanoidins was found to increase while maturing, with a continuous evolution of the starting composition profile detected in the cooked must, allowing molecular sizes to increase beyond 2,000 kDa after 12 years of maturing. The overall quantity of these biopolymers is an extensive property of TBV and it represents the degree of overall chemical change induced by heat during grape juice cooking and successively during vinegar storage. For this reason, the accumulation of these biopolymers, together with some statistically defined ranges of molecular size distribution, have been proposed as markers for the physical maturing time of TBV. Furthermore, the overall distribution profile could be used as a fingerprint for TBV composition with regard to the molecular size. It is expected that exploitation of high-resolution SEC columns, collecting different melanoidin fractions, and investigating melanoidin properties and interactions as a function of their molecular size, will help simplify the study of the mechanisms of melanoidin accumulation and how their distribution may influence the final quality of TBV. It is easy to recognise the high-throughput potential of this methodological approach, especially concerning the rational design of more complex final characteristics of TBV, providing breakthroughs in vinegar science and marketing.

3.4.2 Colour and Spectrum Absorbance

As mentioned above, the colour of TBV changes, while maturing, from yellow brown to dark brown to nearly black, as a result of the accumulation of chromophore-labelled melanoidins (Falcone and Giudici 2008). The authors observed that at least four average sizes of melanoidins contribute differently to the colour and proposed a Brown Index (BI), formulated as the ratio between the absorbance at 420 nm of TBV and the absorbance of the cooked grape must, as a scale for the increase in brown colour over time.

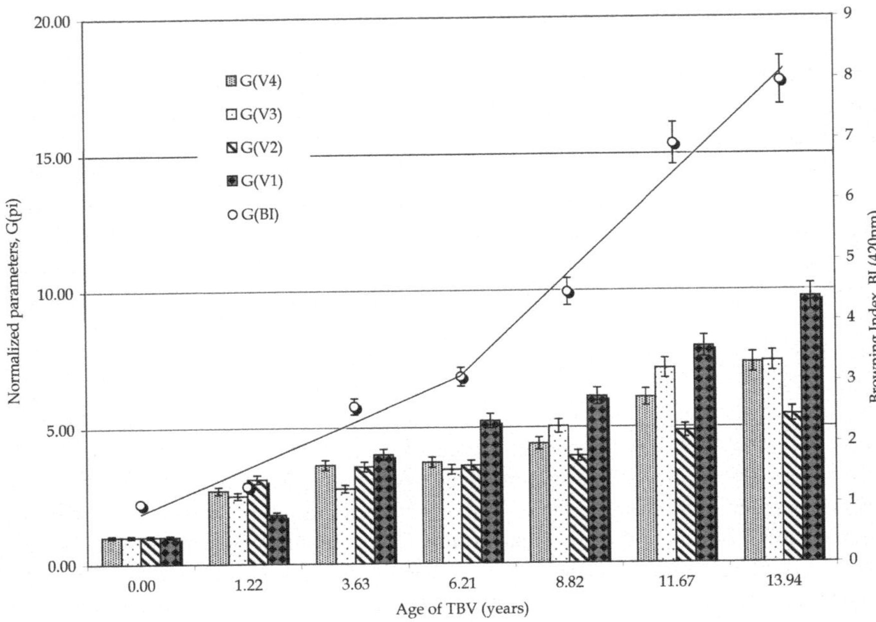

Fig. 3.3 Time course of the browning index (absorbance at 420 nm) in the TBV samples. *Bars* represent the relative concentration of the brown melanoidin population. All data were normalised with respect to their corresponding values in cooked must and referred to as G(V1), G(V2), G(V3), G(V4), and G(BI). From Falcone and Giudici (2008)

$$\mathrm{BI} = \left\{ \frac{A_{[\mathrm{RT_{AV}}(n)]}}{A_{\mathrm{CKM}}} \right\}_{420\,\mathrm{nm}} \tag{3.1}$$

In Eq. (3.1), $\mathrm{RT_{AV}}(n)$ is the average residence time of the ageing vinegar (AV) calculated according to the Giudici and Rinaldi model (2007).

Figure 3.3 shows the BI time course for vinegar obtained from cooked grape must with a barrel residence time (RT) up to 14 years. The relative fraction of coloured melanoidins is also shown in the same figure. The development of browning in TBV samples follows a two-stage determining kinetics, with the second stage starting in about the sixth year of maturing. This finding agrees with visual observations of colour.

The complete absorption profile of vinegar samples matured up to 14 years, as a function of the radiation wavelength within the ultraviolet-visible field (from 200 to 700 nm), is shown in Fig. 3.4.

The spectra exhibited a well-defined absorption peak at 280 nm and featureless end absorptions as the wavelength decreases, in agreement with findings for other naturally occurring melanoidins. There was evidence for a second absorption peak around the 218 nm wavelength from the 10th year of maturing, whereas this peak exhibited a bathochromic effect over the remaining maturing period. This

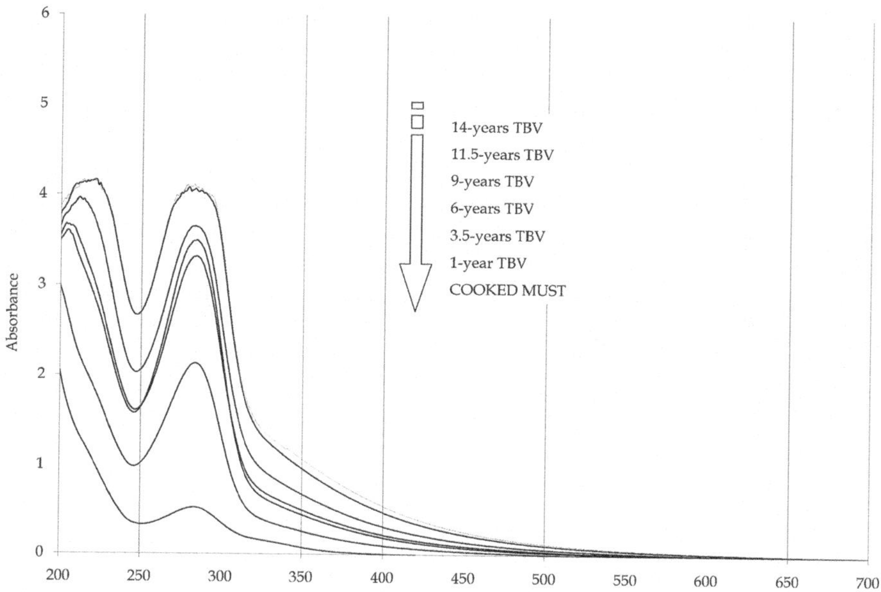

Fig. 3.4 UV-visible spectra of melanoidins extracted from TBV samples with increasing maturing time, from zero (cooked must) to 14 years

wavelength shift was assumed to be due to an increase in the extent of chromophore conjugation in melanoidins, resulting in a copigmentation effect. A similar spectrum profile was registered for a synthetic solution of melanoidins after preparation following a modified version of the method of Shore et al. (1984). In their work, the authors proved that the interaction between a simple amino acid and two simple monosaccharides, here glucose and fructose, leads to chromophore and molecular size distributions very closely resembling those naturally occurring in the cooked must.

3.4.3 Rheological Properties

It has been proved that the rheological properties of TBV are related to the vinegar's viscosity. From a physical perspective, glucose and fructose act as structure-promoting compounds in TBV, both sugars contributing to the vinegar's viscosity, with the glucose following a power law and fructose a more complex path expressing a minimum of its structuring ability (Falcone et al. 2006). Aqueous solutions containing glucose, fructose, and acetic acid, in a range of concentrations simulating a typical TBV, exhibited a Newtonian viscosity, while samples of TBV exhibited a Newtonian viscosity of two or more orders of magnitude greater than the model system. The authors observed that the sum of the non-investigated minor constituents exerts greater structuring influence in comparison to the main vinegar constituents, largely glucose and fructose, allowing the vinegar's viscosity to reach over two orders of magnitude greater than that of the TBV model system. The authors concluded that the minor constituents in TBV cannot be ignored when evaluating the relationships between composition and viscosity. The higher levels of viscosity observed for TBV were attributed to the strengthening of solute interaction caused by the presence of biopolymers formed during the cooking of grape juice and accumulating during maturing (Falcone and Giudici 2008). The authors suggested that some physical properties, including colligative properties, refractive index, density, specific heat capacity, and melt and solution viscosity, as well as rheological properties, like the viscoelastic properties (time-dependent mechanical properties), are the cumulative result of solute interactions, to which several classes of polydispersed constituents may contribute differently. Furthermore, the authors hypothesised the free radical-mediated mechanism as the preferred pathway for solute polymerisation, and the polydispersion index of the biopolymers was found to be above 2 on average.

3.4.4 TBV Composition Over the Last Three Decades

The option of producing cooked musts containing more than 50 % in weight of soluble solids is currently of great interest to TBV producers. However, this practice has affected the composition of TBV over time, because the high solute concentration tends to inhibit alcohol oxidation by AAB and increase selective pressure inside the medium. This phenomenon favours the most able-to-grow osmophilic yeasts, with an increase in the glucose to fructose ratio, and generating a driving force for glucose crystallisation.

Over the last three decades, the sugar content and the overall composition of TBV in general have undergone profound changes. The considerable increase in soluble solids means that the TBV produced nowadays is less pungent and sweeter than in the past. This new taste seems to better match consumer preferences and is favoured by professional tasting panels who award sweeter vinegars with higher sensory scores. The relationship between composition, maturing time, and sensorial properties of TBV has recently been investigated by various authors, applying a statistical approach to the data (Chinnici et al. 2003; Cocchi et al. 2002, 2006; Plessi et al. 2006; Sanarico et al. 2002; Giordano et al. 2003; Masino et al. 2008). Unfortunately, different conclusions were reached and the relationship between age, composition, and sensorial properties remains an intriguing unresolved question. The failure to resolve this issue was partly due to the sample of TBV examined, which were not validated with a reliable age assessment procedure.

3.5 How to Discriminate Between Balsamics

As discussed above, all the balsamic products share important traits and intended uses, but they differ widely in ingredients and production procedures, which can influence prices enormously (see Sect. 4.1). As a result, consumer choice is driven by marketing strategies that claim numerous specific attributes (age, history, tradition), but the condiments are rarely described on the basis of chemical composition or other objective qualitative data.

The grouping of balsamic products into four different categories proposed above is an attempt to clarify the issue, with the ultimate goal being the definition of objective methods and procedures to differentiate any one kind from the others. It is obvious that the chemical composition of balsamic condiments is extremely heterogeneous due to the different types and relative percentages of the various ingredients, blending systems, applied technologies, flavourings, thickeners, emulsifiers, and preservatives. This variability is high both between groups and within the group themselves, and it is easy to find BC samples very similar to TBV, or BV resembling BC and BVM, with prices from a few euros to more than 100 euros for a small bottle. In this chaos of products and prices, consumers are completely disoriented.

A first distinction can be made between vinegars on one hand and sauces, dressings, and other balsamic products on the other hand. BV and BVM are true vinegars, within a specific class of condiments regulated on a national basis (see Chap. 1). All the other balsamic products are foodstuffs without specific composition defined by law, but they must satisfy the general regulations for the safe production of food. Under Italian legislation, TBV do not satisfy the minimum acidity requirement for wine vinegar and so they are treated as condiments.

This chapter sets out a procedure based on a few simple objective parameters to discriminate between four groups of balsamics.

3.5.1 Sugars

The sugars in balsamic products originate from different raw materials and ingredients. In BVM and TBV, grape must is the only source of sugars. Grape must contains between 14 and 24 % in weight of the monosaccharides glucose and fructose, in a ratio close to 1:1. In the other balsamic products, sugars can be added freely, usually sucrose and glucose. Reducing sugars can be quantified easily using the colorimetric/titrimetric Fehling method. In addition to its sensorial effects, a high concentration of sugar improves the syrupy texture. Generic balsamic vinegars have relatively high fluidity and they usually have a more pungent acidity and a lower sugar content than the other categories. Balsamic condiments intended for food dressings exhibit a wide range of sweetness because producers are free to add sweeteners and thickeners to achieve the required taste and texture.

The amount of sugar in BVM depends on the concentration of the must, the acidity of the wine vinegar, and the mixing proportions. Vinegar regulations require at least 6 % of total acidity. Considering that concentrated musts for BVM do not contain more than 1 % of acidity, and using the most expensive wine vinegar with 14 % of acidity, the minimum dilution of the concentrated must with wine vinegar is approximately 38 %. However, wine vinegars for blending generally have lower acidity, so the dilution of the must is higher. For example, using vinegar with 12 % acidity, the minimum dilution of the must is about 45 %.

In other words, the concentration of sugar required in the final product determines the concentration of must to be used, since concentration after blending is not allowed in the product regulations. An increase in concentration due to water loss occurs, to a minimal extent, only in matured BVM.

In order to obtain BVM with a high relative density, the concentration of the cooked or concentrated must has to be extremely high, together with a highly acidic vinegar. The minimum specific gravity imposed by the BVM regulations is 1.24, corresponding to about 51 % (w/w) of sugars. Minimal dilution of this must results in BVM with 37 % sugar. In high-density BVM, sugars exceed 50 %, so the initial must needs to have a concentration close to 70 %.

The highly concentrated musts for BVM production cannot be used for TBV, since such high sugar concentration would be disastrous. Firstly, cooking at atmospheric pressure (as required in the TBV regulations) would produce excessive caramelisation. Secondly, alcoholic and acetic acid fermentation would be inhibited. Finally, if the initial content of soluble solids is too high, the physical stability of the TBV would be at risk during the long maturing period.

The cooked must for TBV generally has a sugar concentration between 30 and 40 %. This value decreases due to alcoholic fermentation and direct oxidation of glucose by AAB. Subsequent maturing tends to progressively increase the concentration of sugars up to 50 %. The limit of 50 % of sugars was observed in a large number of TBV samples.

Moreover, it should be noted that a high quantity of reducing sugars is not sufficient to explain the high specific gravity of TBV samples. The discrepancy between the predicted and observed relative density values indicates the presence of solutes (derived from sugar) undetectable by titration with Fehling's reagent. This is an indirect confirmation of the profound changes that the sugars undergo during the slow maturing process. Similar observations can be made for the Brix value, which is completely different from the predicted value based on the concentration of sugar, especially for extensively matured products.

3.5.2 Brix and DeltaZ

The Brix value is a measurement of the percentage by weight of sugars, using a refractometer. The Brix scale refers to standard aqueous solutions of sucrose, which is a disaccharide composed of one unit of glucose and one of fructose. For mixtures of sugars similar to the standard solutions, such as grape juice and other fruit juices, the refractometer provides values that are comparable with those obtained from titration with Fehling's method. As the complexity of the mixture increases, other substances interfere with the measurement, and so the correspondence between the Brix degree and the actual quantity of reducing sugars progressively declines.

For BV, BC, and BVM, which mainly contain simple sugars, the brix value is slightly higher than the sugar content. In TBV, the Brix value is always above 60. Due to the large number of compounds that can cause interference, and the pronounced degradation of simple sugars, the adimensional difference between Brix value and reducing sugars by weight (DeltaZ) is very marked. In high-quality TBV samples, DeltaZ is greater than 20 and it can even exceed 30 units (Lemmetti and Giudici 2011). DeltaZ quantifies the rate of degradation, modification, and polymerisation of sugars. Clearly, this phenomenon only occurs after a very long maturing period. The comparison between Brix value and sugar content also provides some indication of the risk of solidification (Falcone 2010; Lemmetti and Giudici 2011).

3.5.3 *Acidity*

The total acidity from acid–base titration is the sum of the contributions of volatile and non-volatile acids. The grapes from the Modena and Reggio Emilia regions provide musts for wine making with acidity levels between 6 and 15 g/l, with the Lambrusco cultivar producing the highest acidity and therefore the highest content of malic acid. Musts for vinegar production generally come from low-grade grapes, by second pressing of over-ripened grapes with low acidity. Therefore, the total acidity of products made from these juices and wine vinegar is almost exclusively due to the volatile acidity of the acetic acid. The fixed acidity of BVM rarely exceeds 1 % by volume. For BV without specific limitations on ingredients, fixed acidity can be higher due to the addition of food grade acids.

With regard to TBV, several factors are responsible for the generally high fixed acidity. Firstly, Trebbiano and Lambrusco are the most commonly used varieties for TBV. In order to preserve the acidity of the must, it is strongly recommended to harvest early in order to maintain high level of malic acid with a positive effect on the acids/sugars ratio. Furthermore, the yeasts involved in the fermentation of cooked must help improve the fixed acidity of TBV due to the production of succinic acid (some species of yeast produce this acid more than others). AAB influence the fixed acidity of TBV due to the production of gluconic acid. In TBV, the fixed acidity plays a major role in sensory quality because it makes it possible to obtain vinegar with acidity as high as 10 % while keeping the pungency low. In comparison with other balsamic products, the fixed acidity of TBV is quite high. Analyses of TBV samples indicate that the minimum value for fixed acidity is 1.6 % by weight, corresponding to at least 2.4 % by volume (Lemmetti and Giudici 2011).

Therefore, fixed acidity is another significant parameter for discriminating matured vinegars (made from fermented must) from other products made using concentrated must and wine vinegar.

3.5.4 pH

pH is proportional to the hydrogen ion concentration and is measured using an electrode sensitive to this ion. pH depends on the amount of acids in solution and their dissociation constant. More correctly, pH is the measurement of the *activity* of hydrogen ions. This depends on the ionic strength of the solution, related to the presence of ionic species that modify the availability of ions $(H_2O)_nH^+$. Apart from extreme cases, the activity of an ionic species decreases in concentrated solution, as ionic strength increases. The titratable acidity is therefore linked to the pH by a complex series of factors: apart from the nature and quantity of acids, other substances influence pH measurement. It is well known, for example, that in wines the pH increases as the concentration of potassium increases, because the higher ionic strength lowers the activity in solution.

Comparing pH measurements in various balsamic products, an interesting behaviour was observed. In BVM, as in other non-matured vinegars, the response of the pH meter is proportional to the titratable acidity and the pH settles at values greater than 2.8. In TBV, there is no clear relationship between pH and acidity. The pH value is always less than 2.8, and can be as low as 2.3. The explanation for such behaviour is highly complex and beyond the scope of this book. On one hand, it can be due to the presence of chemical species with acidic constants higher than that of acetic acid, which is the main source of acidity in other balsamic products. On the other hand, it would be expected that concentrated mixtures, such as TBV, exhibited high ionic strength, reduced activity, and higher pH. Probably, the measurement of pH of balsamic vinegar with high relative density is affected by instrumental error because the conditions of use of the pH meter are far outside the ideal range. Regardless of the actual causes of the observed phenomenon, pH can be used as a parameter to discriminate between balsamic products.

Table 3.6 shows the categories of balsamic products in relation to the above-mentioned discriminating parameters.

According to the discriminating weight of each parameter identified, a decision tree was proposed (Fig. 3.5). It outlines a sequence of analyses to identify real matured balsamic vinegars. The DeltaZ value should be sufficient to identify TBV, but fixed acidity and pH can be used for further confirmation. However, differentiation between the industrial products (BVM and other balsamic condiments) is not as easy.

TBV is the only product in the balsamic family that can be identified from a few physical parameters within the capacity of any basic analytical laboratory. This is a result of the extended maturing period, which allows the formation of compounds from the degradation of sugars that no longer react with Fehling's reagent. Other

Table 3.6 The set of parameters used to discriminate between balsamic vinegars

Balsamic product	Reducing sugars (Z)	Brix	DeltaZ = Brix − Z	pH	Total acidity (g of acetic acid/ 100 ml)	Fixed acidity (g of acetic acid/ 100 ml)
Generic balsamic vinegar (BV)	Variable	$\approx Z$	<10	Variable	>6	Variable
Balsamic glazes, sauces, and condiments (BC)	Variable	$\approx Z$	<10	Variable	Variable	Variable
Balsamic Vinegar of Modena (BVM)	Variable	$\approx Z$	<10	>2.80	>6	<1
Traditional Balsamic Vinegars (TBV)	≤50	≥60	≥20	≤2.80	>4.5[a] (Modena) >5 (Reggio Emilia)	>2.4

[a]g/100 g

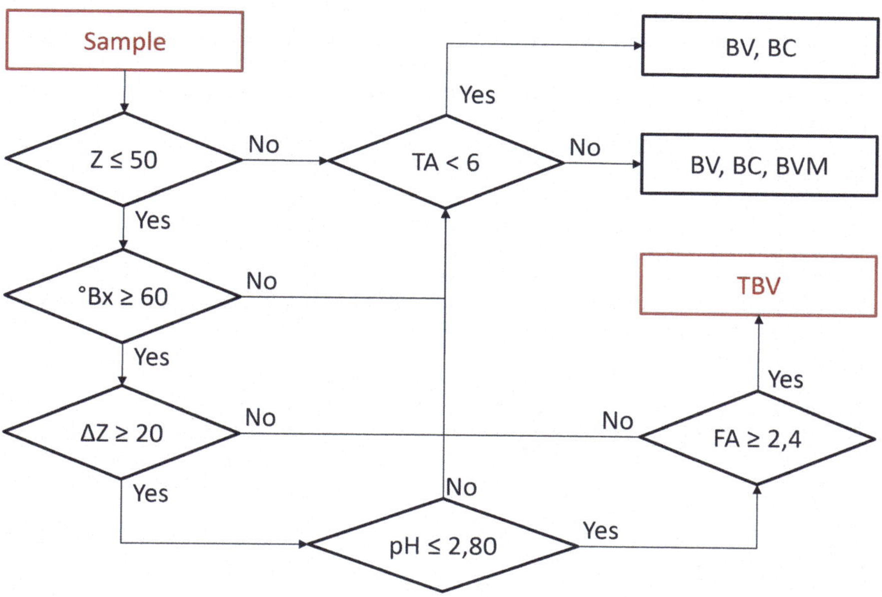

Fig. 3.5 Decision tree for the identification of real matured balsamic vinegar. Starting from an unknown sample, five basic determinations are required

balsamic products, BVM and other condiments, are not easily identified. This implies that most of these products are very similar to each other in regard to composition, maturing, and functional properties, regardless of the protected geographical status.

References

Bononi M, Tateo F (2009) Determination of furan by headspace solid-phase micro-extraction gas chromatography–mass spectrometry in balsamic vinegars of Modena. J Food Compos Anal 22:79–82

Chinnici F, Masino F, Antonelli A (2003) Determination of furanic compounds in traditional balsamic vinegars by ion-exclusion liquid chromatography and diode array detection. J Chromatogr Sci 41:305–310

Cocchi M, Lambertini P, Manzini D, Marchetti A, Ulrici A (2002) Determination of carboxylic acids in vinegars and in Aceto Balsamico Tradizionale di Modena by HPLC and GC methods. J Agric Food Chem 50:5255–5261

Cocchi M, Durante C, Grandi M, Lambertini P, Manzini D, Marchetti A (2006) Simultaneous determination of sugars and organic acids in aged vinegars and chemiometric data analysis. Talanta 69:1166–1175

Consonni R, Cagliani LR, Benevelli F, Spraul M, Humpfer E, Stocchero M (2008) NMR and chemometric methods: a powerful combination for characterization of balsamic and traditional balsamic vinegar of Modena. Anal Chim Acta 611:31–40

Del Signore A (2001) Chemiometric analysis and chemical and physical parameters of traditional balsamic vinegars from Modena. J Commodity Sci 40(1):33–48

EFSA, European Food Safety Authority Scientific Opinion on the re-evaluation of caramel colours (E 150 a,b,c,d) as food additives. Retrieved 15th Sept 2014

Falcone PM (2010) Crystallization and jamming in the traditional balsamic vinegar. Food Res Int 43(8):2217–2220

Falcone PM, Giudici P (2008) Molecular size and molecular weight distribution impacting traditional balsamic vinegar ageing. J Agric Food Chem 56(16):7057–7066

Falcone PM, Giudici P (2010) Aceto Balsamico Tradizionale: indicatori di invecchiamento e del rischio di solidificazione. Industrie delle Bevande 226:19–36

Falcone PM, Chillo S, Giudici P, Del Nobile MA (2006) Measuring rheological properties for applications in assessment of traditional balsamic vinegar: description and preliminary evaluation of a model. J Food Eng 80:234–240

Falcone PM, Boselli E, Frega NG (2011) Structure-composition relationships of the traditional balsamic vinegar close to jamming transition. Food Res Int 44:1613–1619

Giordano L, Calabrese R, Davoli E, Rotilio D (2003) Quantitative analysis of 2-furfural and 5-methylfurfural in different Italian vinegars by headspace solid-phase microextraction coupled to gas chromatography–mass spectrometry using isotope dilution. J Chromatogr A 1017:141–149

Giudici P (1993) Acido gluconico: Criterio di genuinità nell'aceto balsamico tradizionale. Industrie delle Bevande 124:123–125

Giudici P, Rinaldi G (2007) A theoretical model to predict the age of traditional balsamic vinegar. J Food Eng 82:121–127

Giudici P, Gullo M, Solieri L, Falcone PM (2009) Technological and microbiological aspects of traditional balsamic vinegar and their influence on quality and sensorial properties. Adv Food Nutr Res 58:137–182

Janzowski C, Glaab V, Samimi E et al (2000) 5-hydroxymethylfurfural: assessment of mutagenicity, DNA-damaging potential and reactivity towards cellular glutathione. Food Chem Toxicol 38:801–809

Lemmetti F, Giudici P (2011) Semplici parametri chimico-fisici per riconoscere gli aceti balsamici invecchiati. Industrie delle Bevande 235:12–24

Masino F, Chinnici F, Bendinic A, Montevecchia G, Antonelli A (2008) A study on relationships among chemical, physical, and qualitative assessment in traditional balsamic vinegar. Food Chem 106(1):90–95

Muratore G, Licciardello F, Restuccia C, Puglisi ML, Giudici P (2006) Role of different factors affecting the formation of 5-hydroxymethyl-2-furancarboxaldehyde in heated grape must. J Agric Food Chem 54:860–863

Natera R, Castro R, de Valme García-Moreno M, Hernández MJ, García-Barroso C (2003) Chemometric studies of vinegars from different raw materials and processes of production. J Agric Food Chem 51(11):3345–3351

O'Brian J, Morrisey PA (1989) Nutritional and toxicological aspects of the Maillard browning reaction in foods. Crit Rev Food Sci 28:211–248

Plessi M, Bertelli D, Miglietta F (2006) Extraction and identification by GC-Ms of phenolic acids in traditional balsamic vinegar from Modena. J Food Compos Anal 19:49–54

Regulation (EC) No 510/2006 on the protection of geographical indications and designations of origin for agricultural products and foodstuffs. Offic J Eur Communities 2006 (49):1–12

Regulation (EC) No 583/2009 entering a name in the register of protected designations of origin and protected geographical indications [Aceto Balsamico di Modena (PGI)]. Offic J Eur Communities 2009 (52):7–11

Regulation (EC) No 813/2000 supplementing the Annex to Commission Regulation (EC) No 1107/96 on the registration of geographical indications and designations of origin under the procedure laid down in Article 17 of Regulation (EEC) No 2081/92. Offic J Eur Communities 2000 (43):5–6

Sanarico D, Antonelli A, Giudici P (2002) Modello di simulazione di concentrazione dell'aceto balsamico tradizionale. Industrie delle Bevande 182:26–32

Shore M, Broughton NW, Dutton JV, Sissons A (1984) Factors affecting white sugar colour. Sugar Technol Rev 12:1–99

Theobald A, Miller A, Anklam E (1998) Determination of 5-hydroxymethylfurfural in vinegar samples by HPLC. J Agric Food Chem 46:1850–1854

Verzelloni E, Tagliazucchi D, Conte A (2007) Relationship between the antioxidant properties and the phenolic and flavonoid content in traditional balsamic vinegar. Food Chem 105:564–571

Zeppa G, Giordano M, Gerbi V, Meglioli G (2002) Characterization of volatiles compounds in three acetification batteries used for the production of 'Aceto Balsamico Tradizionale di Reggio-Emilia'. Ital J Food Sci 3:247–266

Chapter 4
Balsamic Production: Raw Materials and Processes

4.1 Balsamic Vinegar Manufacturing Processes

The processes for manufacturing balsamic vinegars can differ remarkably between producers, essentially in response to the intended use of the products, production cost, raw materials, consumer preferences, legal conformity, and very importantly the producers' market strategies. A well-known commercial principle states that the more profit is shared along the distribution chain, the more products are sold. This is also true for vinegars and currently a huge amount of balsamic vinegar of low production cost is sold at a reasonable price, sharing the profit along the distribution chain. In contrast, products like TBV, with its high production costs, are very hard items to sell for the opposite reason: they are genuinely expensive and, proportionally, less profit is shared along the distribution chain.

The balsamic manufacturing processes are limited to a few specific stages, since ingredients are often bought partially or totally processed, including concentrated grape must, cooked grape must, fruit juices, vinegar, glucose syrups, and other ingredients. Table 4.1 reports the basic single operations, ingredients, and production costs of the different types of balsamics as reported in Chap. 3. The costs are estimated on the basis of average prices (current year on the Italian market) of the main ingredient used and without considering packaging and all the other marketing costs like boxes, labels, and advertising. The range of balsamic selling prices are the minimum and maximum prices, expressed by litre of product, retrieved from retail and web pages consulted in the year 2014.

The processes of manufacturing balsamic vinegar can be divided into a few stages or operations, some common to all balsamics and serving the same purpose, such as improving colour, viscosity, and density, which are common traits of all balsamics, together with their sweet and sour taste (see Chap. 3).

The first stage is the formulation with choice of ingredients, raw materials, and/or their proportions in the mixture, which is particularly important for BC,

P. Giudici et al., *Balsamic Vinegars*, DOI 10.1007/978-3-319-13758-2_4

Table 4.1 Raw materials, process, production cost, and selling prices of balsamic vinegars and condiments

Balsamic type	Basic raw materials	Must concentration and cooking	Fermentations	Blending	Production cost (€/kg)	Price (€/l)	Maturation
BC	Grape must, wine vinegar, syrups, sugars, starch, thickeners, flavours, other food ingredients	Concentrated or cooked grape must can be used. A further cooking step can follow	Typically none Vinegar is used as ingredient	Yes	Up to 4.00	16.00–24.00 Sometimes exceptional prices—1,000 € and more!	Occasionally (not mandatory)
BV	Grape must, wine vinegar, caramel colour (E150d)	Concentrated or cooked grape must can be used. A further cooking step can follow	Any fermentation, vinegar is used as ingredient	Yes	Up to 4.00	3.00–10.00	Occasionally
BVM	Grape must, wine vinegar, caramel colour (E150d)	Concentrated or cooked grape must can be used. A further cooking step can follow prior to the final blending step	Any fermentation, vinegar is used as ingredient	Yes	Up to 5.00	10.00–50.00 Sometimes exceptional prices 1,000 € and more !	At least 60 days
TBV	Cooked grape must	Cooking provides the required concentration	Alcoholic and acetic fermentation occur spontaneously	No	150.00–300.00	400.00–900.00	At least 12 years

BV, and BVM, since some of these products can be manufactured directly in this single first stage.

Two important basic operations involve heat and mass transfer for concentration and cooking of the grape must. For the production of TBV, the two operations are carried out together and in a traditional open cauldron, while for the other balsamics there are several options. The biological steps of alcoholic and acetic acid fermentation are critical steps for TBV, due to the presence of high levels of sugars, and are traditionally carried out in static and uncontrolled conditions (see Chap. 5).

For the other balsamics (BC, BV, BVM) the process is different, and producers have various options. In the first stage, the grape must is available in its cooked (caramelised) or concentrated variants. Legally, these are two separate ingredients, each with its own definition (OIV 2014). The producer chooses one by preference, or might decide that both are required for the finished product.

For the subsequent formulation of BC, BV, and BVM, many options are available, summarised below:

- Simple blend: the grape must (concentrated, cooked, or concentrated and thermally treated) is blended with other ingredients (wine vinegar, caramel, syrups, additives, etc.) - BC, BV, or BVM.
- Partial mixing followed by cooking and final blending: e.g. to obtain a high density intermediate product which is blended in different ratios with wine vinegar to obtain the desired profile - BC, BV, or BVM.
- Blending and cooking: only for BC or BV, not for BVM which for legal reasons cannot be modified after final blending.

Unlike TBV production, the wine vinegar used as an ingredient for BC, BV, and BVM is obtained from submerged fermentation using very sophisticated equipment, with precise control of temperature, dissolved oxygen, alcohol, and acetic acid contents.

4.2 Raw Materials and/or Ingredients

The distinction between raw material and ingredient is arbitrary since the same component could be both, e.g.: grape must could be an ingredient if used directly in the formulation of a balsamic or a raw material to be transformed into cooked must or vinegar. The major ingredients used for the production and/or formulation of balsamics are described below.

4.2.1 Vinegars

Acetic acid for alimentary use must be of agricultural origin (the type obtained from wood and coal is strictly forbidden in food) and is the result of a double fermentation, first alcoholic and then acetic. The final content of acetic acid is dependent on the alcohol level and the initial sugar content of the raw materials.

Fresh fruit contains from 4 to 20 % sugars, dried fruit and concentrates can have levels above 70 %, while the amount of sugar contained in a "mash" of cereals depends on the dilution and the yield of the process of hydrolysis of starch.

In general, the content of acetic acid increases from fruit vinegars (3 to 6 % w/v), to wine (8 to 14 %), and finally the spirit vinegar from the alcohol of agricultural origin obtained by distillation of fermented crops can be as high as 18–20 % (Ebner et al. 1996; Gullo et al. 2014). The crops frequently used for spirit vinegar vary by country and include sugarcane, sugar beet, fermented fruit, and, after an enzymatic or chemical hydrolysis process, corn, barley, rice, and other cereals.

The different plant species are distinguished by being highly selective in the setting of stable isotopes of carbon and hydrogen, so the isotope ratios of the organic matter of different species can easily be identified by NMR or other isotopic methods. This is based on the differentiation of the relationship between the isotopes of H, C, and O due to different metabolism in the setting of type C_3 and C_4 plants or relative to the natural abundance of isotopic species. These methods (and other traditional approaches), applied to sugars, alcohols, or CO_2, are currently in use for wine, and some have recently been developed for the analysis of wine vinegars (OIV 2001).

The botanical origin of the acetic acid does not pose any nutritional or safety concerns, but it has a strong economic and technological relevance. The first is simply a question of price, while the second relates to the concentration of acetic acid in the vinegar, since high concentrations are very useful for creating high density balsamics. For example, the specific gravity of BVM depends on the density of the must, the acidity of the wine vinegar, and the mixing proportions. BVM regulations require at least 6 % total acidity. Considering that concentrated musts for BVM do not have more than 1 % acidity, using the most expensive wine vinegar with 14 % acidity, the minimum dilution of the concentrated must with wine vinegar is around 38 %.

However, wine vinegars usually have lower acidity, so the dilution of the must is higher. For example, using vinegar with 12 % acidity, the minimum dilution is about 45 %.

In other words, the density required in the final product determines the concentration of the must and vinegar to be used. The minimum specific gravity imposed by the BVM regulations is 1.24, representing about 51 % (w/w) sugars. Minimum dilution of this must gives a BVM with 37 % sugars. In high density BVM, sugars exceed 50 %, so the initial must needs a concentration close to 70 %. Moreover, when ingredients are mixed, the sugar content of the BVM does not undergo substantial changes.

The botanical origins of the alcohol used for spirit vinegars are typically fruits in tropical countries, but quantitatively more important are sugar beet, cereals (maize, oats, rice), sugarcane, potatoes, and other crops. With the exclusion of TBV and BVM, which should be manufactured using only juice from grapes, the other balsamics can be made with raw material from any agricultural crops. Recently, it has been reported that the practice of using crops other than grapes is also occasionally practiced for BVM (FCR Magazine 2011; Stiftung Warentest Test 2011), but this does not conform with the PGI product regulation.

4.2.2 *Grape Musts*

The definitions of vitivinicultural products are included in the first part of the International Code of Oenological Practices (OIV 2014), and grape must is defined as the liquid product obtained from fresh grapes, whether spontaneously or by physical processes such as crushing, removing stems from grape berries or crushed grapes, draining, and pressing.

Preserved grape must is fresh grape must in which alcoholic fermentation has been prevented by one of the following oenological procedures: sulphiting, addition of diethyl carbonate, or sorbic acid (illegal in some countries). A small quantity of endogenous ethanol in grape musts is tolerated, with a limit of 1 % (v/v).

To achieve biologically stable grape must at room temperature, the concentration of sulphites should be close to 2,000 ppm. Before using this preserved grape must to make wine or for any other purpose, the sulphites must be removed. The sulphite removal techniques are physical with stripping in a vapour stream or under vacuum, or chemical with the addition of oxygen peroxide. The oxidation of sulphites drastically increases the amount of sulphates in the must. During the concentration process sulphites are again removed and partially oxidised into sulphates. As a consequence, desulphited musts usually have a higher level of sulphates compared to fresh musts.

It is well known and documented that the characteristics and composition of a must depend on the grape variety, climate, agronomical techniques, and time of harvesting. It is less obvious what the ideal must composition is for balsamic production. For TBV, the characteristics of the must are described and discussed in Chap. 6, with particular attention to the sugar and titratable acidity ratio. The question arises as to whether there are general rules valid for all balsamics, or each kind of balsamic needs its own specific grape juice composition.

Before answering this question, it should be noted that for balsamics with protected status (BVM and TBV), grapes are the only raw material legally permitted, and no other ingredients can be used to optimise the sugar/acidity ratio. In contrast, for BC and BV, food grade sugars and acids can be freely used for the formulation of these products. Furthermore, sugars from corn, sugar beet, or other agricultural crops are less expensive than sugars from grapes, the same applying to food grade acids like malic, tartaric, lactic, and citric. Essentially, the composition of the grape juices is an important factor only for balsamic vinegars with official geographical indication.

4.2.3 Concentrated Grape Must

Concentrated grape must (CGM) is neither fermented nor caramelised and is obtained from the partial dehydration of grape must or of preserved grape must according to procedures accepted by the OIV, such that its density at 20 °C is not less than 1.24 g/ml. Fresh or desulphited must can be concentrated by several methods, the most common being evaporation under vacuum at low temperatures, but reverse osmosis and cryoconcentration could also offer practical advantages for prestigious products. The price of CGM is proportional to the amount of soluble solids (usually expressed in °Bx).

Concentrated grape must can be directly used in the formulation of balsamics, and/or with caramelisation by heat treatment, to improve colour and sensorial properties. The composition of CGM is strictly dependent on the level of water removed and on the initial must composition. The main organic constituents of CGM are glucose and fructose in a ratio close to 1, and tartaric and malic acids. During the concentration process, a significant amount of tartaric acid precipitates as tartrate salts. The ratio between sugar and titratable acidity is an important quality parameter for CGM, because it is better that the desired total acidity of a vinegar is achieved with a lower level of volatile acidity, mainly acetic acid (see Chap. 7). For oenological purposes, there is special rectified CGM, in which the acid fraction is removed by treatment with cation exchange resins (Barbieri and Rossi 1980). These musts can also be used for balsamic production to increase the sugar content without affecting the acidity.

4.2.4 Cooked Grape Must

Cooked grape must or caramelised grape must (CKM), according to the definition of the OIV, is a non-fermented product, obtained by the partial dehydration by direct heating of grape must or of grape must preserved according to procedures accepted by OIV, such that its density at 20 °C is not less than 1.3 g/ml.

4.2.5 Glucose Syrup

Glucose syrup is obtained by chemical or enzymatic hydrolysis of starch from several crops like maize, rice, and potatoes. Starch corn is the main raw material in the USA, where it is called corn syrup, while rice starch is widespread in China and the Far East.

Glucose syrup contains different amounts of sugars depending on its intended use. Over 90 % glucose serves for industrial fermentation, while syrups used to make confectionery contain varying amounts of glucose, maltose, and higher oligosaccharides, depending on the grade, with the amount of glucose varying from 10 to 43 % (Dziedzic and Kearsley 1995; Jackson 1995; Norman and Hendriksen 2001).

The composition of glucose syrup is an important issue for food and balsamic vinegar production, since the ratio between the constituents, such as sugars, oligosaccharides, and high molecular weight compounds (partially hydrolysed starch), determines the rheological properties of the syrup. In BC and BV, glucose syrup is widely used to increase the viscosity of products and, indirectly, other sensorial properties, for example, sweetness, which is a complex perception related to the concentration of simple sugars but also viscosity. Similarly, in beer and wine the molecules that stimulate the sensation of sweetness are the same that contribute to the "body" or "smoothness", due of their ability to increase the viscosity (see Chap. 7).

There are several kinds of glucose syrup on the market specifically prepared for different purposes: thickeners, sweeteners, humectants, texture softeners, volume additives, to prevent crystallisation of sugar, and as flavour enhancers.

The syrups are roughly categorised according to their dextrose equivalent (DE). The further hydrolysis proceeds, the more reducing sugars are produced, and the higher the DE. In the past, glucose syrups were manufactured by acid hydrolysis of corn starch at high temperature and pressure. Higher DE syrups made by acid hydrolysis tend to have a bitter taste and a dark colour, due to the production of hydroxymethylfurfural and other by-products (Hull 2010).

At present, glucose syrup is mainly produced by first adding the enzyme α-amylase to a mixture of corn starch and water. The enzyme breaks the starch into oligosaccharides, which are then broken into glucose molecules by adding the enzyme glucoamylase (Steinbüchel and Rhee 2005).The gelification of starch is a fundamental step to increase the effectiveness of enzymatic treatments. Figures 4.1 and 4.2 show equipment for steaming sticky rice before enzymatic treatments with a raw mixture of da-qu, the fungal mycelium preparation for Chinese rice vinegar production (Chen et al. 2009).

4.2.6 Caramel

Caramel is one of the oldest and most widely used food colours. Its colour ranges from pale yellow to amber to dark brown. It is manufactured by heating carbohydrates, either alone or in the presence of acids, alkalis, and/or salts (Kamuf et al. 2003). The Codex Alimentarius recognises four types of caramel, each with its own INS and E number, differing by the reactants (sulphite and/or ammonium, both or none) used during manufacture and the intended and/or authorised use: caramel E150a mainly used for Whiskey; E150b for Cognac, Sherry, and also for some vinegars; E150c for beer, sauces, and confectionary; E150d for balsamic vinegars, and other acidic environments such as colas.

Fig. 4.1 Equipment for steaming sticky rice: rice steaming basket

Caramel colour is globally approved for use in food, but application and intake restrictions vary by country. Together with its primary colouring function, caramel has rheological properties and also protects bottled beverage aromas against oxidation. Two parameters are important to assess the quality of caramel: the colour intensity (CI) and the colour tone (Grover 1968). The first is defined as the absorbance (A) at 610 nm of a 0.1 % (w/v) caramel solution normalised to the total solids (TS):

$$\mathrm{CI} = A \times 100/\mathrm{TS}.$$

The second parameter (Linner Hue Index) is a function of the absorbance of light at 510 and 610 nm and is a measure of the colour hue or red characteristics of the caramel colour (Linner 1970).

Caramels, especially those prepared using ammonium salts (E150c and E150d), are object of controversial safety concerns due to the presence of 4-Methylimidazole (MEI). Some public safety organisations (e.g. FDA; EFSA) consider caramel safe at the levels generally used in food and consistent with good manufacturing practice.

Fig. 4.2 Equipment for steaming sticky rice: vapour distribution

4.2.7 Other Ingredients or Additives for High Viscosity

As already noted, the *balsamic family* of products also includes condiments characterised by a cream-like consistency, typically made with a liquid part composed of grape must, balsamic vinegar of Modena, and the addition of ingredients that increase density and viscosity.

One of the most commonly chosen ingredients to increase viscosity is the abovementioned glucose syrup, much appreciated by technologists for its neutral taste and low price, but to obtain the desired viscosity it is always necessary to add more specific ingredients or additives. There is only a limited choice of options because very few substances can withstand the low pH (always around 3) and the highly oxidative environment of this class of products.

Modified corn starch (E1422) is the most widely used by producers for the purpose, followed by functional hydrocolloids like xanthan gum (E415) or guar gum also called guaran (E413). Other ingredients like native starches, pectins, citrus fibre, carragenins, agar, and numerous others have found limited applications in the balsamic field, mainly because the majority of them are not stable under the highly acidic conditions of the finished products. They are prone to degradation by acid hydrolysation and return to a completely liquid state, or even the opposite, becoming too reactive and causing complete jellification or solidification.

4.3 Processing Operations for Balsamics

4.3.1 Evaporation and Concentration

Ideally, evaporation and concentration are different processes. Evaporation involves a simultaneous transport of heat and matter, mainly water removed in the form of steam. Concentration consists in removing water from solutions or dispersions to obtain concentrated products, with a water content over 20 %. There are several technologies to achieve the desired concentration: evaporation, reverse osmosis, cryoconcentration, ultrafiltration, dialysis, and electrodialysis. Evaporation is a concentration technique by which the product to be concentrated is brought to boiling temperature, and the water is eliminated in the form of steam.

In food preparation, evaporation is generally conducted at low pressure to reduce the boiling temperature and preserve the characteristics of the food, while for balsamics the degradation of sugars at high temperatures (caramelisation) is often considered a positive effect and there is a specific operational cooking stage. Industrial concentration of grape juice and glucose syrups is achieved using special equipment, single or multi-effect evaporators, and under strictly controlled pressure and temperature to minimise sugar degradation. There is substantial literature on the topic (Asghari and Yoshida 2006; Barbieri and Rossi 1980; Di Leo 1988) to which readers are referred. In the traditional method of concentrating grape must in an open cauldron, evaporation, concentration, and cooking are carried out simultaneously. This process is described in detail in the following paragraph.

4.3.2 Cooking

Cooking is an extended process for some traditional Italian foods and beverages, including traditional balsamic vinegars and some special liquors known as *vino cotto* (cooked wine). Cooked grape must is also produced in Spain for sweet wine production (Rivero-Pérez et al. 2002). The general use of cooked must in European countries is regulated by International laws (Regulation EC 1493/99).

In the specific case of balsamics, the cooking of CGM, concentrated glucose syrups, and any other sugary syrups has two main objectives which are improving the sensorial properties and colour. It is well known that thermal treatment induces formation of high molecular size biopolymers, some of them leading to the formation of brown compounds, the so-called "melanoidins" (Falcone and Giudici 2008). These compounds are thought to have significant effects on the final quality and consumer acceptance of widely consumed dietary goods (e.g. coffee, cocoa, bread, malt, and honey), also thanks to their claimed antioxidant properties (Borrelli et al. 2002; Verzelloni et al. 2007), antimicrobial activity (Del Castillo et al. 2006), antihypertensive properties, prebiotic activities (Borrelli and Fogliano 2005), browning properties (Bozkurt et al. 1998), and foam stability (D'Agostina et al. 2004).

As noted above, the traditional cooking method by direct heating in an open cauldron causes water evaporation, concentration of the syrup/must, and caramelisation of sugars. This traditional method has been completely abandoned in industrial BV and BC production, but persists for home-made BC, TBV, and to a limited extent for BVM. Generally, the devices used for the concentration of grape must are also used for cooking in a two-step operation: concentration at low temperature and under vacuum and then cooking.

An open cauldron is still utilised for cooking the concentrated must, usually with capacity ranging from 5 to 30 tons. This equipment and methodology mimics the traditional cooking method, with a further increase in density, and simultaneous caramelisation.

The best industrial cooking method is the use of pasteurisation devices (Fig. 4.3), which offer the desired level of cooking with control of time and temperature and, more importantly, the possibility to control the level of harmful unwanted compounds that can form during heat treatment. It is well known that heating carbohydrates leads to the formation of some potential toxic compounds, including methylglyoxal, acrylamide, and furan derivatives, such as 5-hydroxymethylfurfural (HMF). Various studies have shown that HMF acts as a cytotoxic, genotoxic, and tumorigenic agent (O'Brian and Morrisey 1989; Janzowski et al. 2000; Zhang et al. 1993).

The EFSA Panel on Contaminants in the Food Chain elaborated a scientific report reviewing currently available data on methods of analysis, occurrence, formation, and furans exposure toxicity and concluded that the currently limited availability of foods occurrence data does not permit a valid dietary exposure assessment. A call for more information on the occurrence of furans in cooked foods has thus been issued. EU regulations impose limits on the occurrence of HMF in some products: for example, concentrated musts may contain a maximum of 25 mg/Kg of total sugars.

A systemic study of the kinetics of HMF formation during grape must cooking evaluated the individual contributions of glucose and fructose concentration, water activity, and pH level, i.e. all the properties were treated as independent variables. The rate of HMF formation under different sugar concentrations is a function of water activity and to a lesser extent of sugar concentration and pH value (Muratore et al. 2005). However, HMF and methylglyoxal act as intermediate chemicals, and their concentration and lifetime are related to the initial reducing sugar concentration and the extent of sugar degradation.

Gogus et al. (1998) studied some model systems to determine the reactivity of the major sugars (fructose and glucose) and amino acids (glutamine and arginine), observing the accumulation of HMF in grape juice, and brown pigment formation

Fig. 4.3 A small pasteurisation device used for cooking grape must

(BPF) at 55, 65, and 75 °C over 10 days at pH 3.5. The authors observed that fructose is more reactive than glucose in both HMF and BPF accumulation at all temperatures, and that with fructose as substrate, glutamine was more reactive than arginine, but with glucose, arginine was found to be more reactive than glutamine under the same conditions.

Concentrated solutions of glucose convert partially in to a complex mixture of reversion products including fructose, mannose, and oligosaccharides (De Man 1999; Van Dam et al. 1986). A reversion mixture including isomaltose, gentiobiose, cellobiose, and maltose has been isolated in thermal-acid treatment of D- glucose (Thompson et al. 1954). In the presence of hydrochloric acid at 98 °C, glucose forms extensively di-, tri-, and tetrasaccharides, whereas D-fructose showed no formation of such reversion products (Silberman 1960).

4.3.3 *Evaporation, Concentration, and Cooking of Grape Must by the Traditional Single Step Operation*

The vessel used for cooking acts as an unsteady-state batch reactor in which water evaporation is the main physical factor, and sugar degradation is the main chemical factor, driving cooking simultaneously. Working at atmospheric pressure, water evaporation from the must surface is driven by the relative humidity of ambient air, which is not controlled. It is widely accepted that water evaporation involves simultaneous heat and mass transfer and depends on the amount of heat supplied, required to vaporise the water and to remove vapour from the vicinity of the surface from which evaporation takes place. Although water evaporation is the most evident effect of cooking, heating grape must causes the formation of fouling materials, mainly foams and natural colloids (Figs. 4.4 and 4.5).

Cooking rapidly stops all enzymatic browning reactions caused by polyphenol oxidase which is known to occur in fresh grape juices. It then promotes bulk discolouring and clarification, probably due to the thermal deactivation of browning enzymes and, finally, progressive non-enzymatic browning (Fig. 4.6).

4.3.3.1 Chemical Changes

The chemical changes that occur during the simultaneous operation of evaporation, concentration, and cooking are the same as those described in the previous paragraph on cooking. What changes significantly between cooking at constant volume and in traditional open cauldrons are the respective kinetics of formation and, not least, the predictability and control of the final result. The time-dependent accumulation of solutes follows different kinetics, leading to the supposition that chemical potentials arising inside the cooked must are reactant specific. After the temperature is quenched, cooked must behaves as an out-of-equilibrium system, reducing all chemical potentials.

A double role of cooking can be hypothesised in sugar degradation, especially during the advanced stages with water evaporation and solute concentration. Glucose and fructose, in particular, are expected to contribute to the lowering of free water molecules due to their high hygroscopic capacity. As a consequence, the proton-catalysed step (Belitz and Grosch 1999) involved in acid-catalysed sugar degradation is continuously advanced throughout cooking, with an increase in the amount of degradation products. This hypothesis is supported by results from some studies focused on the formation of 5-HMF in sugar solutions (Livingston, 1952; Eichner and Karel 1972) as well as in model systems simulating grape must

Fig. 4.4 Cauldron device for cooking grape must

Fig. 4.5 Removal of fouling formed during the heating of grape must

(Muratore et al. 2005). The latter authors observed that the formation of 5-HMF at fixed and limiting sugar concentrations was a function of water activity and to a lesser extent of initial sugar concentration and pH value. In particular, an inverse linear relationship was found between 5-HMF concentration and water activity varying from 0.779 to 0.889, while an inverse power-type relationship was found relative to the pH value, varying from 2.9 to 3.2.

From the perspective of safety, the quantity of furans, including 5-HMF, is extremely important in cooked must production. 5-HMF starts to form and accumulate after a lag period, but fractional conversion increases with cooking time following to a linear trend. The linear trend indicates a zero-order kinetic of 5-HMF formation (Falcone et al. 2010). The kinetics of HMF formation was also investigated by other studies, which reported that this reaction fitted zero-order, half-order (Bozkurt et al. 1998), first-order, and second-order kinetic models (Shallenberger and Mattick 1983).

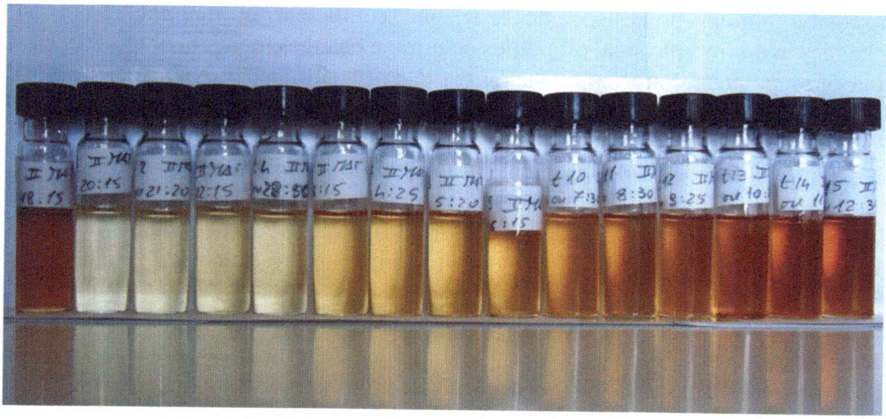

Fig. 4.6 The effect of cooking on the colour of grape must. Samples were collected during traditional cooking in open vessels with direct heating. The heating time increases from *left* to *right*

In addition to the evolution of the main constituents, changes in some minor compounds play an important role in achieving the quality properties of TBV, in particular newly formed compounds of large molecular size, above 500 kDa, both labelled with uncoloured and brown chromophores, deriving from polymerisation reactions. These compounds are the limit-of-equilibrium state of the polymerisation reactions: they are first activated upon cooking and then progress more slowly during the maturing of balsamics (Falcone and Giudici 2008).

Size-exclusion chromatography analysis of cooked musts from the Trebbiano cultivar confirmed the presence of refractive compounds of molecular size broadly dispersed among the glucose and fructose molecules (Falcone and Giudici, 2008).

Differences in total phenolic compounds during cooking were explained on the basis of the strong tendency of these compounds to suffer polymerisation reactions. It is widely known that the polymerization of phenolic compounds is especially favoured by acidic pH and reduction in water content (Fulcrand et al. 2006). Moreover, the presence of aldehydes in cooked musts such as HMF may also catalyse the polymerisation reaction of flavonoids. However, when the degree of polymerisation exceeds a critical value, the increased molecular complexity promotes a decrease in antioxidant capacity, probably because of steric hindrance, which reduces the availability of the phenolic rings for the horseradish peroxidase activity. An extended polymerisation of natural phenolic compounds could explain their negative conversion in the advanced stages of cooking (Falcone et al. 2010).

4.3.3.2 Physical Changes

It is very evident during heating that grape juice undergoes other profound changes related to solute interactions, generally affecting juice density, refractive index, viscosity, boiling point, specific heat, and coefficient of thermal expansion. Changes in thermo-physical properties determine changes in the rheological behaviour of grape juice, in turn influencing the rate of heat transfer (Rubio et al. 1998).

Generally, clarified juice concentrates exhibit a Newtonian behaviour (Rao et al. 1984; Saravacos 1970; Schwartz and Costell 1986), although some authors have noted a minor pseudoplasticity in the flow of grape must at concentrations above 55 °Bx. This was attributed to the presence of some insoluble solids, mostly pectins and tartrates (Moressi and Spinosi 1984). However, other authors observed the juice concentrates behaving as Newtonian fluids even at high soluble solid concentrations of 60–70 °Bx (Barbieri and Rossi 1980; Di Leo 1988; Rao et al. 1984; Schwartz and Costell 1986). Falcone and Giudici (2008) suggested that changes in both thermo-physical and rheological properties of grape juice for TBV production depend on the heat-induced formation of high molecular size biopolymers. These authors also analysed the effects of glucose, fructose, and acetic acid on the shear viscosity and concluded that rheological measurements could be a simple and reliable method for rapid and objective evaluation of TBV quality before performing sensory analysis. The instrumental measurement of viscosity may have practical value for optimising vinegar properties by designing production methods able to influence the residual sugars, organic acids, and physical density ratios (Falcone et al. 2007).

4.4 The Balsamic Production Process

The ingredients and processing procedure described previously are partially shared with the balsamic categories reported in Chap. 3. The basic processes and ingredients are described and discussed below, with an attempt to highlight the similarities and differences. It is worth underlining that all the balsamics share some basic characteristics: colour, viscosity, sweet and sour taste, and often the intended use. These characteristics can be obtained from different procedures and ingredients, and production costs have a strong influence on applied options. For example, the high viscosity of a BC can be achieved easily with the addition of thickeners, such as modified starch, or it can be the result of extended maturation as with TBV.

4.4.1 Traditional Balsamic Vinegars from Modena and Reggio Emilia

The TBV process starts from freshly squeezed grapes and finishes with sensory evaluation of the vinegar by an official panel. Sensorial evaluation is mandatory in order to achieve PDO status, and each lot of vinegar must receive a positive score in order to use the registered names "Traditional Balsamic Vinegar of Modena"

(TBVMO) or "traditional Balsamic Vinegar of Reggio Emilia" (TBVRE) on the label. The sensorial score awarded is used to rank TBV into different commercial classes: two for TBVMO and three for TBVRE.

The specific regulations for TBVMO allow adding "Extra Vecchio" to the official designation when the product is particularly excellent and is claimed to be matured for 25 years or more. However, neither the definition of *maturing* nor the methods for its objective evaluation are specified (see Chap. 6). The ranking is made exclusively with panel-tasting tests, the effectiveness of which for age assessment are clearly inadequate.

There are not many processing stages between the grape must and sensory evaluation: must cooking, acetification of the cooked must, and maturation. During these stages numerous major physical, biological, and chemical transformations occur.

The grape juice is cooked in open vessels heated directly over a fire for 12–24 h. The heating drives three important processes: water evaporation, concentration, and chemical transformations, due to Maillard reactions and caramelisation (see previous paragraphs). The temperature is usually kept a few °C below boiling point, and heating is stopped when the solute concentration of the CKM is between 30 and 50 °Bx. The product regulations require grape must with a minimum concentration of soluble solids of 15 °Bx and after cooking at least 30 °Bx for TBVRE, with an upper limit of 50 °Brix for both the PDO's.

The CKM is transferred, still hot, into vessels that are immediately sealed hermetically, permitting preservation without any additional treatment. Alternatively, the CKM can be left to cool and poured into a special large barrel called a *badessa*, a word probably derived from the monastery term *Abadess*, underlining the important role of this barrel. The badessa is the main site of the biological transformations, where fermentation and acetification of the CKM take place.

It was generally believed that alcoholic and acetic fermentation occurred in the badessa in a commensalistic relationship (Sacchetti 1974). However, the yeast and acetic acid bacteria require different conditions to grow, and they have a reciprocal negative influence on each other. What happens in the badessa depends on the composition of the medium: low volatile acidity privileges alcoholic fermentation, after which the alcohol becomes the substrate for acetification. If, as expected, AAB develop, new acetic acid is formed and alcoholic fermentation is completely repressed.

To avoid the arrest of alcoholic fermentation, part of the vinegar is removed, and new cooked must is added to decrease the volatile acidity. The process is repeated every year, but unfortunately irregular fermentations are frequent, either alcoholic or acetic. Even more seriously, neither of the two might occur, with a negative influence on the sensorial properties of the TBV. In particular, the absence of acetic acid fermentation gives TBV with low volatile and titratable acidity. To avoid this problem one option is to add fresh vinegar as a starter culture. The addition of vinegar rapidly inhibits alcoholic fermentation completely and no more alcohol is formed, giving origin to a cyclical process that requires new vinegar every year.

Fig. 4.7 Barrel with a river stone used to cover the bunghole. The granular sand released by the solubilisation of the river stone carbonate is evident on the cloth

Recently, the fermentation stage has been radically revised, with the two fermentations conducted in two different vessels (see Chap. 5).

The vinegar obtained in the badessa is used to fill a set of barrels, generally from 5 to 7, of different capacity and made from different wood varieties (oak, chestnut, cherry, juniper, mulberry, acacia). The barrels are partially filled with vinegar from the badessa, with a head space of a few cm left between the surface of the vinegar and the top of the barrel, and their bungholes are covered with a towel. In the past the bungholes were covered with river stones, and carbonates from the stones were dissolved by the vapours (acetic acid) from the barrels, with silicate released in the form of sand grains (Fig. 4.7).

The barrels are conventionally numbered from right to left, from smallest to largest: the TBV is tapped off from barrel n° 1 while the last barrel receives new vinegar from the badessa. Only a portion of the vinegar is removed from barrel 1, which is then refilled with vinegar from the next barrel. This procedure is called *rincalzo* and is repeated for each barrel in the set, the last one receiving new cooked must. The purpose is to replace both the vinegar drawn off and lost by evaporation, as well as the vinegar transferred from barrel to barrel. With this procedure, every barrel contains a blend of vinegars of different *ages* (see Chap. 6).

From a legal point of view, TBV must be aged at least for 12 years, but no objective procedure has been specified for age assessment and vinegar classification. This is a very weak point for the assurance of quality and authenticity of TBV because there is a legal requirement but no criteria or procedure to demonstrate it.

The barrel set plays an important role in the ageing process. It is composed of casks made of different wood varieties. In the past, the different woods were considered important to increase the sensorial properties of TBV, but the authors are not aware of any scientific evidence to support this. Furthermore, the raw taste of the different woods is considered a defect for TBV, and new barrels have to be "deflavoured" before use. To achieve this, new barrels are filled with wine vinegar and left to stand for at least 6 months, after which the vinegar is removed and the barrels are ready for TBV maturation.

The real role of wood barrels in TBV ageing is related to their gas permeability (oxygen and water vapour) and molecular selectivity based on molecular size. The wood acts as a semi-permeable membrane and selects the molecules able to cross the wood on the basis of their steric dimensions. In general, the structure of wood is sufficient to stop all molecules with dimensions greater than water, such as alcohols, esters, and acetic acid (all molecules with sensorial significance). In contrast, water can freely pass through wood in the vapour state (see Chap. 5).

4.5 The Common Balsamics: A Redeemed Tradition

As discussed in Chap. 2, the history of balsamic vinegars is very long and complex, but it is impossible, at the current state of knowledge, to establish exactly how they evolved through time. What does appear clear is that at some point in history, balsamic vinegars started to be part of daily life when the so-called *Duke's vinegar* became famous outside of castles and palaces, with the result that common people started to manufacture balsamic vinegars at home.

Scholars like Rangone and Giacobazzi have demonstrated that many different qualitative types of vinegars already existed in the eighteenth century, and there are numerous references to the difference between a quality defined as *common* (*comune* in Italian), for which the main ingredient was wine vinegar, and superior qualities called *da padrone* (*master's grade*), which were produced with higher proportions of grape must (Rangone and Giacobazzi 2011).

These observations demonstrate that, alongside the highly valued small scale production of "noble" vinegar, characterised by very long maturation times and incredibly expensive, there has always been a tradition of balsamic vinegars made from a blend of wine vinegar and grape must, in a wide range of qualities. This variety served different uses and budgets, just as large and even small companies produce certified PGI protected balsamic vinegar of Modena in several commercial variants today. The basic requirement is that the key features of balsamic vinegars are achieved: their pungent and sweet taste, dark colour, and complex sensorial profile.

4.6 Industrial Balsamic Vinegars

In the scenario outlined above, the industrial BVM manufactured today in modern plants, can be seen as a continuity of the common variant, descended from a single stem of the countless evolutionary branches of Modena-style vinegars. The production of BVM typically starts with a blend of grape must and wine vinegar, with the optional addition of caramel to adjust the colour saturation to its required intensity. The caramel used for industrial applications is not the culinary pan-heated browned sugar, but a colouring additive manufactured in a specific process, in the presence of chemical catalysts and with dedicated equipment. As regards the wine vinegar and grape must, these have already been discussed in the previous paragraphs of this chapter.

The product regulations for *Aceto Balsamico di Modena* PGI (Regulation EC 583/2009) prescribe adding "a quantity of vinegar aged for at least 10 years", even if the actual "quantity" has never been further specified. Predictably, many producers interpret this requirement in the most advantageous sense, adding a truly symbolic amount of this expensive 10-year-old ingredient. However, some producers add a more significant amount of extensively matured vinegar to enhance and distinguish their high bracket products.

Physical blending is achieved simply by mixing in tanks with the aid of rotors, turning paddles, or sometimes air inflation to generate convection. The blending time depends mainly on volume and overall product viscosity: the higher the viscosity, the longer the fluid mass needs to be moved to become homogeneous. Small batches (e.g. 500 l) of a basic quality might require just a few minutes, while large batches of dense viscous products (e.g. 100,000 l) might require an hour and a half.

After blending, the prepared mass of product must undergo a mandatory period of maturation or ageing inside a wooden container (barrels, casks, vats, etc.). According to the product regulations, there are two different levels of ageing in wood for BVM:

- 60 days (2 months): this is the legal minimum for the product to be certified as a PGI *Aceto Balsamico di Modena*
- 3 years: this is the minimum ageing period for a product to be certified as *Matured* (*Invecchiato*, in Italian)

After the ageing period, the product must be certified as PGI by an authorised control body on behalf of the Italian government. The producer must submit an official request for testing and authorisation to the control body, which dispatches auditors to the production site to take samples of each batch ready for certification and sealing the containers in which it is stored. The submitted BVM is analysed and tested. After a few days, the control body gives a response, issuing a certificate if the product is approved or informing the manufacturer of non-conformity if the product is rejected.

Approved BVM is certified as PGI *Aceto Balsamico di Modena* and from this time onwards it can be bottled, used as an ingredient (e.g. in a compound product) or sold in bulk.

For packaged products, there is a specific labelling authorisation procedure that must be followed. Labels, seals, and any kind of packaging material intended for a retail product must be submitted to the BVM Protection Association (*Consorzio Tutela Aceto Balsamico di Modena*), who is responsible for approving or rejecting it. The packaging must bear the legal name *Aceto Balsamico di Modena* along with the wording *Indicazione Geografica Protetta* (protected geographical indication) written in full or as an acronym, in Italian and/or in the language of the country of destination. The name of products certified as *Aceto Balsamico di Modena* IGP cannot be qualified in any way, even in numerical form. Commercially appealing adjectives of any kind, like fine, special, distinguished, classic, superior, or similar cannot be printed on the packaging alongside the protected name. Only the word *matured* (*invecchiato*), without any further additions, can appear, provided that the product has been previously approved and certified as matured.

4.7 Condiments with Balsamic Vinegar

The recent commercial success of BVM has expanded the world market with the appearance of numerous different condiments made with balsamic vinegars or other types of vinegars (wine vinegar, apple cider vinegar, honey vinegar, fruit vinegars, etc.). Many producers have invented and offered the public their own condiments, always maintaining the basic profile and characteristic sweet and sour flavour of balsamic vinegars, but offering an incredible number of products, differing in colour, flavour, texture, and packaging.

The production of these so-called condiments is legally separated from TBV, BVMs, or other vinegars and is regulated by general food legislation. It started in the smaller TBV market, as a way of selling products derived from the same production process as TBV, and retaining similar features, but avoiding the extended maturation times required by the product regulations for the traditional PDO products. However, since the mid-1990s the larger BVM producers have also started to develop condiments using larger and more competitive production processes.

The first industrial scale condiments that were sold on the market were basically colour variants of BVMs, always as blends of wine vinegar and concentrated grape must, or in some cases also glucose syrup. The aim of those condiments was to propose light coloured versions of balsamic vinegar, with a fresher more fruity flavour that would be suitable for cooking but without altering the colour of the culinary preparations, e.g. light coloured sauces, fish, salads, and many others. These condiments are still very popular today in numerous markets around the world and have become part of the standard range of any retailer or distributor of balsamic vinegars. They are normally referred to as *white condiments* or *white balsamic condiments*, even if the term *balsamic* is not used by the PGI certified producers, because it infringes legal and voluntary guidelines derived from PDO and PGI protection. They are available in different densities and colour variants ranging from pale yellow to burgundy red.

A second type of condiment started to multiply at a very impressive rate starting from the early 2000s: glazes or creams, made with BVM. Glazes were invented in the Modena area by BVM producers at the end of the 1990s. The initial aim of these glazes was to imitate the reductions of Balsamic Vinegar of Modena created in the kitchens of restaurants and catering services by cooks. The main property of these condiments is their viscosity and cream-like appearance, which is produced by certain specific ingredients like starches or other stabilisers. The glazes are used as a simple substitute of BVM, as a condiment, or as a topping. They are typically sprinkled on to ice cream, strawberries, and matured cheeses or used to garnish the meat and vegetables of main courses. As condiments, they can be packed in squeezable plastic containers, which are much more convenient and efficient than the glass bottles required for BVM packaging. The turnover of these products is increasing year by year, due to their success with HORECA, but also among retail consumers. Glazes are available in a number of variants and can easily be customised to market demands. They can include added fruit juices, flavours, herbs, cocoa, nuts, spices, mushrooms, and so on. They can be differentiated in both flavours and colours: alongside the classic black BVM glaze, the market also offers white glazes (with wine vinegar), red or violet glazes (e.g. with raspberries, blood oranges, pomegranates), green glazes (with herbs or mint), and so on.

The production process varies widely, because there are many ways to obtain glazes. These range from very simple mechanical mixing processes, to complex temperature-controlled phases of heating and cooling conducted by large companies. The market for glazes and similar products is still expanding and the forecasts are for further growth in the near future, while there remains the as yet little explored market for high viscosity ready-to-use condiments at very affordable prices.

References

Asghari FS, Yoshida H (2006) Acid-catalyzed production of 5-hydroxymethyl furfural from D-fructose in subcritical water. Ind Eng Chem Res 45:2163–2173

Barbieri R, Rossi N (1980) Proprietà fisiche dei mosti d'uva concentrati. Rivista Viticoltura Enologia Conegliano 1:10–18

Belitz HD, Grosch W (1999) Food chemistry, 2nd edn. Springer, Berlin, pp 252–261

Borrelli RC, Fogliano V (2005) Bread crust melanoidins as potential prebiotic ingredients. Mol Nutr Food Res 49:673–678

Borrelli RC, Visconti A, Mennella C et al (2002) Chemical characterization and antioxidant properties of coffee melanoidins. J Agr Food Chem 50:6527–6533

Bozkurt H, Gogus F, Eren S (1998) Nonenzymic browning reactions in boiled grape juice and its models during storage. Food Chem 64:89–93

Chen F, Li L, Qu J, Chen C (2009) Cereal vinegars made by solid-state fermentation in China. In: Solieri L, Giudici P (eds) Vinegars of the world. Springer, Milan, pp 243–259
D'Agostina A, Boschin G, Bacchini F, Arnoldi A (2004) Investigations on the high molecular weight foaming fractions of espresso coffee. J Agr Food Chem 52:7118–7125
De Man JM (1999) Carbohydrates. In: De Man JM (ed) Principles of food chemistry, 3rd edn. Aspen, Gaithesburg, MD, p 163
Del Castillo MD, Ferrigno A, Acampa I et al (2006) In vitro release of angio-tensin converting enzyme inhibitors, peroxyl-radical scavengers and antibacterial compounds by enzymatic hydrolysis of glycated gluten. J Cereal Sci 45(3):327–334
Di Leo F (1988) Caratteristiche fisico-chimiche dei mosti concentrati rettificati. Valutazione gleucometrica. Vignevini 15:43–45
Dziedzic SZ, Kearsley MW (1995) Handbook of starch hydrolysis products and their derivatives. Blackie, London, p 230
Ebner H, Sellmer S, Follmann H (1996) Acetic acid. In: Rehm HJ, Reed G (eds) Biotechnology, vol 6. Wiley-VCH, Weinheim, pp 381–401
EFSA, European Food Safety Authority Scientific opinion on the re-evaluation of caramel colours (E 150 a,b,c,d) as food additives http://www.efsa.europa.eu/en/efsajournal/pub/2004.htm. Accessed July 2014
Eichner K, Karel M (1972) The influence of water content and water activity on the sugar-amino browning reaction in model systems under various conditions. J Agr Food Chem 20:218–223
Falcone PM, Giudici P (2008) Molecular size and molecular weight distribution impacting Traditional Balsamic Vinegar ageing. J Agr Food Chem 56:7057–7066
Falcone PM, Chillo S, Giudici P, Del Nobile MA (2007) Measuring rheological properties for applications in quality assessment of Traditional Balsamic Vinegar: description and preliminary evaluation of a model. J Food Eng 80:234–240
Falcone PM, Tagliazucchi D, Verzelloni E, Giudici P (2010) Sugar conversion induced by the application of heat to grape must. J Agr Food Chem 58:8680–8691
FCR Magazine (2011) Vinagre balsamique. Pas forcément autentique 10:21–23
Fulcrand H, Duenas M, Salas E, Cheynier V (2006) Phenolic reactions during winemaking and aging. Am J Enol Viticult 57:289–297
Gogus F, Bozkurt H, Eren S (1998) Kinetics of Maillard reaction between the major sugars and amino acids of boiled grape juice. Lebensm Wiss Technol 31:196–200
Grover DW (1968) The measurement and character of caramel colour. J Food Technol 3:311–323
Gullo M, Verzelloni E, Canonico M (2014) Aerobic submerged fermentation by acetic acid bacteria for vinegar production: process and biotechnological aspects. Process Biochem 49:1571–1579
Hull P (2010) Glucose syrups: technology and applications. Wiley-Blackwell, Chichester
Jackson EB (1995) Sugar confectionery manufacture. Springer, Berlin, p 132
Janzowski C, Glaab V, Samimi E et al (2000) 5-hydroxymethylfurfural: assessment of mutagenicity, DNA-damaging potential and reactivity towards cellular glutathione. Food Chem Toxicol 38:801–809
Kamuf W, Nixon N, Parker O, Barnum GC (2003) Overview of caramel colors. Cereal Foods World (American Association of Cereal Chemists, Inc) 48(2):64–69
Linner RT (1970) Caramel color: a new method of determining its color hue and tinctorial power. In: Proceedings of the Society of Soft Drink Technologists annual meeting, pp 63–72
Livingston GE (1952) Malic acid-fructose reaction. J Am Chem Soc 75:1342–1344
Moressi M, Spinosi M (1984) Engineering factors in the production of concentrated fruit juices, II, fluid physical properties of grapes. J Food Technol 5:519–533
Muratore G, Licciardello F, Restuccia C et al (2005) Role of different factors affecting the formation of 5-hydroxymethyl-2-furancarboxaldehyde in heated grape must. J Agr Food Chem 54:860–863
Norman BE, Hendriksen HV (2001) Enzymatic preparation of glucose syrup from starch. US Patent 6,287,826, issued 2001-09-11, assigned to Novo Nordisk A/S, Bagsvaerd (DK)

O'Brian J, Morrisey PA (1989) Nutritional and toxicological aspects of the Maillard browning reaction in foods. Crit Rev Food Sci 28:211–248

OIV (2014) International code of oenological practices

OIV, Resolution Oeno 17/2001. Determination by isotope ratio mass spectrometry 13C/12C of wine ethanol or that obtained through the fermentation of musts, concentrated musts or grape sugar

Rangone U, Giacobazzi G (2011) Dall'aquila estense al Tricolore. Tecnograf, Reggio Emilia

Rao MA, Cooley HJ, Vitali AA (1984) Flow properties of concentrated juices at low temperatures. Food Technol 3:113–119

Regulation (EC) No 1493/99 of 17 May 1999 on the common organisation of the market in wine

Regulation (EC) No 583/2009 entering a name in the register of protected designations of origin and protected geographical indications Aceto Balsamico di Modena (PGI)

Rivero-Pérez MD, Pérez-Magariño S, Gonzàles-San Josè ML (2002) Role of melanoidins in sweet wines. Anal Acta 458:169–175

Rubio LA, Munoz E, Carullo CA et al (1998) Evaporadores tubulares de circulacion forzada: capacidad de calir intercambiada en distintas configurac-iones. III Jornadas de Investigación. FCAI–UN Cuyo, Libro de Resumenes

Sacchetti M (1974) Sull'aceto balsamico modenese. Edagricole, Bologna

Saravacos GD (1970) Effect of temperature on viscosity of fruit juices and purees. J Food Sci 35:122–125

Schwartz M, Costell E (1986) Influencia de la temperatura en el comportamiento reologico del azucar de uva (cv, Thompson seedless). Revista de agroquımica y tecnologıa de alimentos 3:365–372

Shallenberger RS, Mattick LR (1983) Relative stability of glucose and fructose at different acid pH. Food Chem 12:159–165

Silberman HC (1960) Reactions of sugars in the presence of acids: a paper chromatographic study. J Org Chem 26:1967–1969

Steinbüchel A, Rhee SK (2005) Polysaccharides and polyamides in the food industry properties, production, and patents. Wiley-VCH, Weinheim

Stiftung Warentest Test (2011) Bluff aus Modena 6:22–28

Thompson A, Anno K, Wolfrom ML, Inatome M (1954) Acid reversion products from D-glucose. J Am Chem Soc 76:1309–1311

Van Dam HE, Kieboom APG, Van Bekkum H (1986) The conversion of fructose and glucose in acidic media: formation of hydroxymethylfurfural. Starch 38:95–101

Verzelloni E, Tagliazucchi D, Conte A (2007) Relationship between the antioxidant properties and the phenolic and flavonoid content in traditional balsamic vinegar. Food Chem 105:564–571

Zhang XM, Chan CC, Stamp D et al (1993) Initiation and promotion of colonic aberrant crypt foci in rats by 5-hydroxymethyl-2-furaldehyde in thermolyzed sucrose. Carcinogenesis 14:773–775

Chapter 5
Traditional Balsamic Vinegar as Coded by Us

There are several descriptions of how to make balsamic vinegars. Benedetti mentions 21 traditional methods that are still practiced on a household level (Benedetti 2004), some of these are quite unusual, others are reasonable but very generic. In Chap. 2 methods are also described for producing balsamics "artificially" and quickly. In general, all the protocols appear more like cooking recipes than industrial processes. Furthermore, the PDO product regulations do not provide enough details to serve as instructions for making TBV. The product regulations describe the permitted grapes, production area, and other information on packaging, but no details are provided as regards the basic production steps: how long the must needs to be cooked, or even more importantly how to ferment or acetify the base wine (BW).

Recently, a procedure was patented (Patent BO2011A000132) perfectly compatible with the product regulations for TBV, which clearly sets out all the steps for TBV production. Ideally, the process is divided into four main stages, with a separate description of the characteristics of the grapes. The grape composition is an important component of TBV quality, because, as for wine, grapes play an important role to the sensory properties of vinegars. In particular, the balance between sweetness, acidity, and pungency is determined by the ratio of titratable acidity/sugars in the grapes.

The four stages (Fig. 5.1) are:

- Cooking of grape must (GM)
- Alcoholic fermentation of cooked must (CKM)
- Acetic fermentation of base wine (BW)
- Maturation

Chemical and physical transformations occur during all stages, beginning with Maillard reactions generated by the heat treatment of the must and ending with the same Maillard reactions in the barrel set. Maillard reactions are accelerated by the high osmotic pressure of the concentrated grape must and concentrated vinegar (Muratore et al. 2006).

P. Giudici et al., *Balsamic Vinegars*, DOI 10.1007/978-3-319-13758-2_5

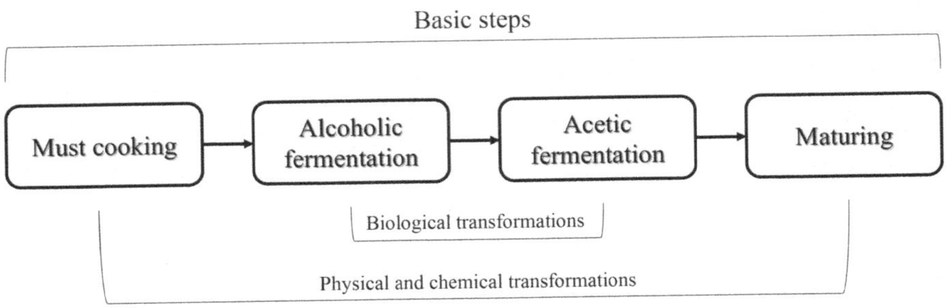

Fig. 5.1 The fundamental production steps of TBV

In the central stages two important microbiological transformations occur: alcoholic and acetic acid fermentations. It is interesting that yeasts and acetic acid bacteria (AAB) cannot develop together significantly, since acetic acid is a strong inhibitor of yeast, and yeast makes the medium anaerobic and unsuitable for AAB growth. The rationalisation of the production of traditional balsamic vinegar into four distinct stages, in particular the separation of alcoholic and acetic acid fermentations, dates back in the early 1990s (Giudici 1990; Giudici et al. 1992). These studies showed that the development of AAB followed alcoholic fermentation and was not simultaneous with yeast growth, because of the strong competition between the two microbial groups.

Before these studies, it was a common belief that alcoholic and acetic fermentation took place in association, in a sort of commensalism non-reproducible in a laboratory (Sacchetti 1974).The two biological processes take more or less 1 year to complete, with the alcoholic fermentation being easier to pilot than the acetic fermentation.

Finally, the base vinegar is transferred in wood barrels where it undergoes slow maturation to develop and concentrate its flavours. The maturing of the base vinegar is the longest stage and occurs inside a set of barrels of different volumes and wood varieties (see Chap. 6). Figure 5.2 shows the overall production process.

5.1 The Grapes

In winemaking, acidity and sugar content are very important for qualitative assessment of grapes and to determine their final use. For good champagnes (Champagne, Cava, Spumante), the required titratable acidity is higher than for wines, and the grapes are usually harvested earlier to preserve the acidity.

The balance between sugar and acids changes as grapes ripen, with sugar increasing as malate decreases. Grape variety and cultivar, agronomical operations, climate, and the degree of grape ripeness strongly influence a wine's composition and sensorial properties. The literature on these topics is extensive, consistent, and widely accepted, so that it is difficult to select and mention the most significant

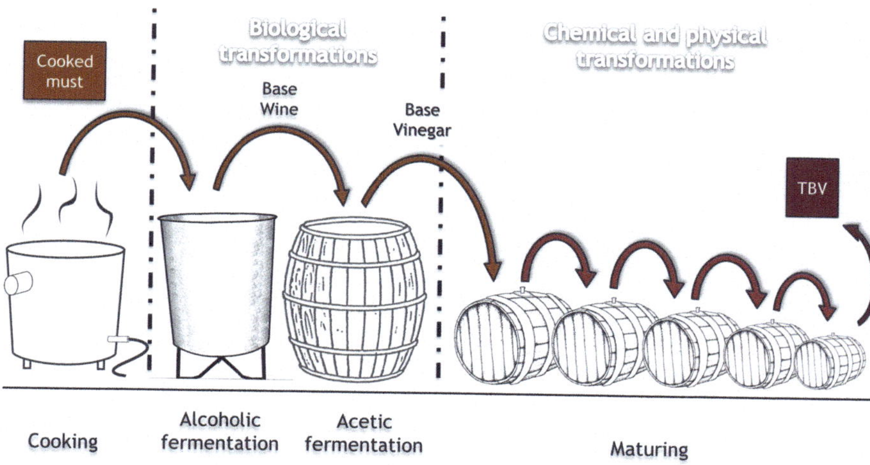

Fig. 5.2 Schematic representation of the Traditional Balsamic Vinegar production process

papers. Fortunately, a good student edition of any oenology and viticulture handbook is adequate to provide a synthetic and comprehensive description of the relationships between grapes, soil/climate, agronomy, wine technology, and wine quality.

Until recently, there were no indications regarding the composition of grapes for TBV, apart from the generic advice to use well-ripened grapes. It was noted that the ratio of sugars and titratable acidity play an important role in consumer preferences (see Chap. 7). According to the product regulations, the grapes permitted for TBV production must come from vineyards of the Modena or Reggio Emilia provinces (Italy), and they should have a minimum sugar content of 15 % w/w. The permitted grape varieties are numerous, both white and red, the most important being Lambrusco (including all varieties and clones), Ancellotta, Trebbiano (including all varieties and clones), Sauvignon, Sgavetta, Berzemino, and Occhio di Gatta.

However, despite the product regulations on grape varieties and their sugar contents being very strict, no scientific studies have been conducted on the quantitative influence of grape composition on TBV quality. Currently, on the basis of sensory evaluation and consumer preferences, the best TBV frequently have very high titratable acidity and low volatile acidity (see Chap. 7), which implies that grapes with high titratable acidity are potentially better than those with low acidity.

Table 5.1 and Fig. 5.3 show data for the macroscopic composition and ripening trends of the most widely used grape varieties for TBV production.

In general, to maintain high acidity, it is necessary to anticipate the harvest, which applies to all varieties. In Lambrusco and Trebbiano cultivars, a high sugar content is generally associated with high titratable acidity throughout the harvest period, while for other local cultivars there is a marked decrease in acidity at the end of ripening. In some cases, the measured values are too low and not compatible with the production of balsamic vinegar.

Table 5.1 Range of qualitative parameters at harvest time of some grape cultivars from Modena and Reggio Emilia: data from the 2010 harvest

Cultivar	Area	Soluble solids (°Bx)	Acidity (g/l tartaric acid)	pH	Ratio sugars/acidity
Lambrusco Salamino	Modena plain	16.52–21.74	9.24–13.95	2.90–3.25	1.20–2.20
	Reggio E. plain	15.18–15.53	12.10–13.67	2.99–3.03	1.14–1.26
Trebbiano Modenese	Modena plain	21.83–22.80	8.23–11.50	3.04–3.28	1.90–2.77
	Modena hills	16.90–20.50	12.90–23.90	2.77–3.25	0.71–1.59
Lambrusco di Sorbara	Modena plain	19.70–22.80	10.38–16.50	2.81–3.74	1.26–1.99
Trebbiano di Spagna	Modena hills	18.50–19.70	9.92–16.80	2.89–3.41	1.17–1.98
	Reggio E. plain	18.50–20.80	9.96–14.80	2.88–3.37	1.37–2.09
Lambrusco Grasparossa	Modena hills	15.30–19.40	11.10–16.50	3.15–3.70	0.93–1.75
	Reggio E. plain	14.50–15.07	8.54–9.07	3.08–3.14	1.60–1.76
	Reggio E. hills	17.96–20.50	5.17–9.01	n.d.	1.99–3.97
	Modena plain	17.90–18.37	6.82–10.92	n.d.	1.66–2.62
Sgavetta	Reggio E. plain	18.60–23.80	6.89–12.70	3.06–3.65	1.62–3.45
Marzemino	Reggio E. plain	15.51–21.00	5.17–9.70	3.05–3.45	2.11–4.06
Ancellotta	Modena plain	18.65–23.60	8.00–9.25	3.09–3.40	2.02–2.93
	Reggio E. plain	15.37–16.83	7.93–10.05	3.14–3.26	1.67–1.94

In the opinion of the authors and based on data on TBV composition and sensory analysis, the following values for the composition of grapes are recommended:

- Titratable acidity greater than 7.5 (expressed in g/l of tartaric acid)
- pH less than 3.2
- Sugar content greater than 16°Bx
- Sugar/acidity ratio between 1.2 and 2.5

The sugar/acidity ratio is calculated by dividing the absolute values of the two variables.

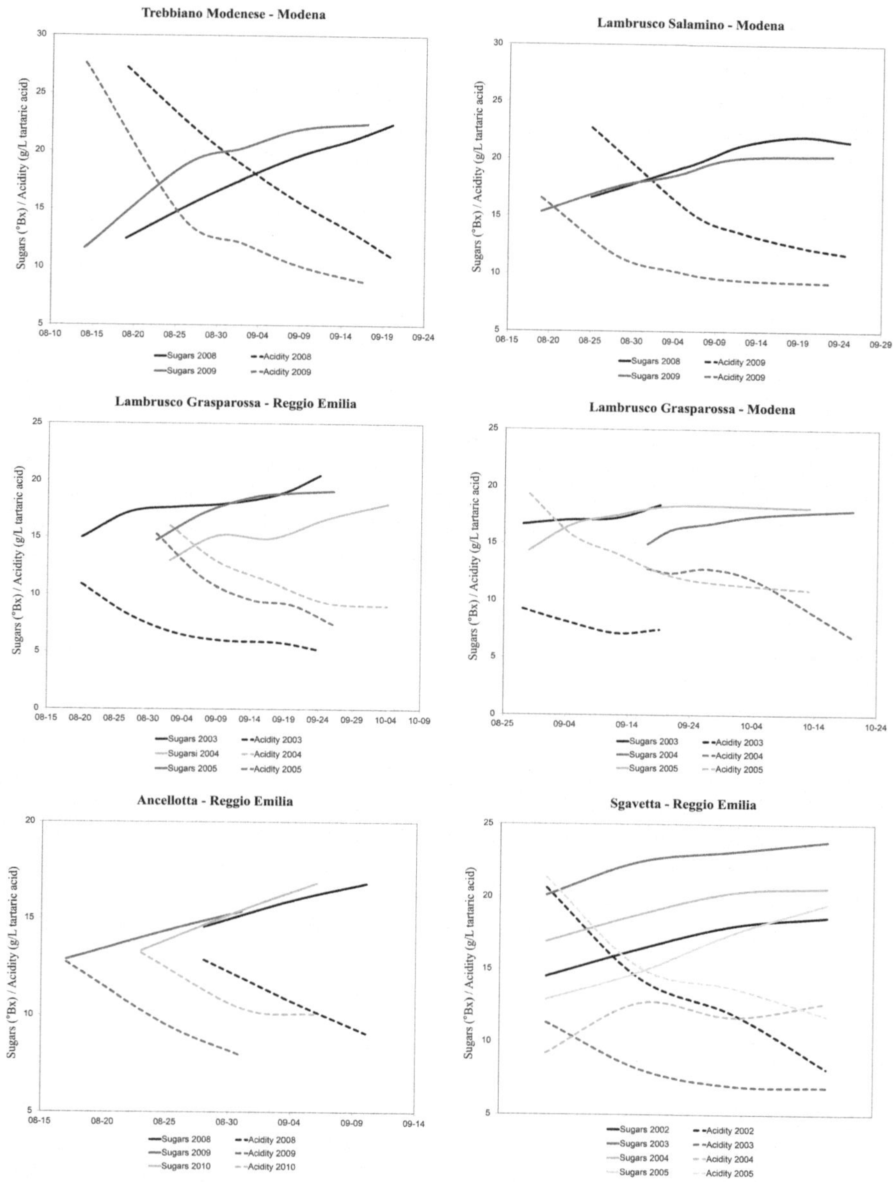

Fig. 5.3 Mean ripening curve of some grape cultivars from the TBV production territory

5.2 The Cooked Must

The GM is the result of the crushing and pressing of grapes. It is well known in oenology that the proportion of juice extracted from grapes makes a big difference to its composition: the greater the pressure on the pomace, the more juice is extracted, and the higher its pH becomes. Individual grapes are not homogeneous in composition, and externally, close to the skin, there is a higher concentration of tartrate salts and a lower content of sugars and acidity. In contrast, the inner pulp is richer in sugars and titratable acidity, while polyphenols are abundant in the pips. Since the sugar/acidity ratio is an important factor, the amount of juice extracted and its composition are important parameters for checking GM quality. In brief, the composition of GM depends on grape variety, time of harvesting, and the crushing and pressing technology used. The greater the must yield (weigh of juice/weigh of grapes as a %), the poorer its composition.

The final composition of cooked must (CKM) is related to the cooking conditions (see Chap. 4), but an important role is also played by the initial GM composition. The first study, relating time of harvesting and GM composition on the effects of cooking, regarded Trebbiano grapes (Falcone and Giudici 2010). Two batches of the Trebbiano cultivar were harvested from the same vineyard 15 days apart , and the musts were cooked in an open stainless-steel cauldron heated over an open flame. The kinetics of the formation or disappearance of key constituents was then monitored for at least 16 h. From an engineering perspective, the vessel behaved like a non-isothermal batch reactor in which the volume of the grape must necessarily decreased while its composition changed profoundly because of chemical reactions. Brix degree, total titratable acidity, acetic acid, malic acid, lactic acid (D- and L-), pH, water activity, 5-HMF, and phenolic and radical-scavenging compounds were proposed as markers of the extent of cooking, with water evaporation and sugar degradation identified as the two main driving factors. Cooking produced opposite conversion of both sugars and acids, with different kinetics between the musts of the short- and long-ripened grapes. More recently a study was carried out on the behaviour of GM from different cultivars and different times of harvesting to assess the influence of GM composition on the final composition of CKM (Table 5.2), focusing particular attention on the sugar/acidity ratio. As previously mentioned, this ratio changes considerably with harvesting time, and the ratio of sugars/titratable acidity after cooking is particularly important. With very few exceptions, the ratio increases after cooking of the must, probably due to the large amount of tartrate, which has low solubility and is in a condition of supersaturate solution in the GM. After cooking and cooling the GM, a huge volume of tartrate is always observed on the bottom of the vessel. Unripened grapes tend to preserve their acidity after cooking, because their acidity is due largely to malic acid, and the salts of this acid are more soluble than tartrates.

Table 5.2 Sugar, titratable acidity, and their ratio before and after cooking of grape must

Sample	Cultivar	Fresh must			Cooking time (h)	Max temp. (°C)	Cooked must		
		Soluble solids (°Bx)	Titratable acidity (g of tartaric acid/l)	Ratio sugars/acidity			Soluble solids (°Bx)	Titratable acidity (g of tartaric acid/l)	Ratio sugars/acidity
GC1	Salamino	14.9	9.0	1.65	13	85	30.8	12.0	2.57
GC2	Salamino	18.0	7.2	2.50	13.5	85	31.6	9.6	3.29
GC3	Salamino	18.0	6.6	2.73	13	85	31.6	12.0	2.63
GC4	Salamino	17.7	7.2	2.46	13	85	30.3	12.0	2.52
GC5	Trebbiano	18.7	6.6	2.83	13	85	32.9	9.6	3.43
GC6	Trebbiano/ Lambrusco	21.3	6.6	3.23	13	85	32.7	9.6	3.41
GC7	Trebbiano	15.8	4.0	3.95	13	85	29.6	5.0	5.92
GC8	Trebbiano	16.6	6.0	2.77	13	85	31.9	9.6	3.32
GC9	Trebbiano	16.5	3.0	5.50	13	85	32.4	6.0	5.40
GC10	Trebbiano	16.9	9.6	1.76	13	85	35.7	12.0	2.97
GU1	Trebbiano	19.8	6.6	3.00	18	81	35.6	9.6	3.71
MR1	Trebbiano	18.0	6.0	3.00	9	90	34.4	9.0	3.82
MR2	Trebbiano	21.4	6.0	3.57	9	90	38.0	9.0	4.22
MR3	Trebbiano	17.2	6.0	2.87	9	90	40.0	9.0	4.44
MR4	Trebbiano	17.8	8.4	2.12	9	90	32.3	9.2	3.51
MR5	Trebbiano/ Grasparossa	18.6	8.4	2.21	9	90	30.1	9.6	3.13
MR3	Trebbiano	17.2	6.0	2.87	16–IIc	90	53.0	9.6	5.52
MD1	Trebbiano Modenese	21.2	9.0	2.35	29	90	31.9	12.0	2.66
MD2	Trebbiano Modenese	20.4	7.2	2.83	19	90	29.4	12.0	2.45

(continued)

Table 5.2 (continued)

Sample	Cultivar	Fresh must			Cooking time (h)	Max temp. (°C)	Cooked must		
		Soluble solids (°Bx)	Titratable acidity (g of tartaric acid/l)	Ratio sugars/ acidity			Soluble solids (°Bx)	Titratable acidity (g of tartaric acid/l)	Ratio sugars/ acidity
MD3	Trebbiano Modenese	20.3	9.6	2.11	24	90	29.2	9.6	3.04
MD4	Trebbiano Modenese	20.6	7.2	2.86	30	90	33.7	9.0	3.74
RAV1	Ancellotta	19.5	5.0	3.90	12	90	32.8	7.5	4.37
SM2	Trebbiano S/M	18.2	9.0	2.02	10	90	28.6	12.0	2.38
SM3	Trebbiano S/M	16.3	8.4	1.94	11	90	29.0	13.2	2.20
SM4	Lambrusco	16.5	9.6	1.72	10	90	26.7	10.8	2.47
LIM1	Ancellotta	22.6	3.0	7.43	14	80	33.6	6.0	5.6
RIG1	Sorbara/ Grasparossa	17.3	7.2	2.40	13	80	31.4	7.2	4.36
RIG2	Trebbiano/ Grasparossa	18.8	6.0	3.13	14	80	33.7	9.0	3.74
RIG3	Ancellotta/ Grasparossa	22.3	8.4	2.65	14	80	33.0	9.6	3.44
FR1	Sorbara	15.8	9.0	1.75	10	75	33.8	18.0	1.88
FR2	Trebbiano Modenese	16.4	9.0	1.82	11	75	31.1	12.0	2.59
FR3	Trebbiano Modenese	16.7	6.6	2.53	10	75	32.0	11.0	2.91

The data shows significant variability also as regards grape cultivars, and those with the higher sugar/acidity ratio, independently of the degree of maturation, are the least suitable for TBV, e.g. Ancellotta

5.3 The Base Wine

The alcoholic fermentation of GM is an ecologically complex process studied since Pasteur's times and, over the last 100 years the dynamics of yeasts during wine fermentation have been described many times, elucidating the role of non-*Saccharomyces* and *Saccharomyces yeasts* (Amerine and Kunkee 1968; Davenport 1974; Fleet and Heard 1993).

In TBV, the role of yeasts was underestimated, but some studies have recently demonstrated that their metabolism is responsible for some of the physical and chemical properties (Landi et al. 2005; Solieri et al. 2006; Solieri and Giudici 2008). The first study on TBV yeast was done by Sacchetti and dates back to the 1930s and later summarised in a book (Sacchetti 1974), in which the author identifies strains belonging to the genus *Zygosaccharomyces* as the predominant TBV yeasts and proposed a commensalistic interaction between yeasts and AAB. In the 1980s, Turtura and co-workers investigated the main TBV-related species (Turtura and Benfenati 1988). They reported the presence of *Z. bailii* and *Z. rouxii*, identified on the basis of morpho-physiological features, such as the ability to grow in 1 % acetic acid concentration. Later, Giudici (1990) also detected the occurrence of *Saccharomycodes ludwigii* strains, together with *Z. rouxii* and *Z. Bailii*. Recently a complex yeast micro-flora, including *Z. bailii*, *Z. rouxii*, *S. ludwigii*, *Z. mellis*, *Z. pseudorouxii*, *Z. bisporus*, and *Z. lentus*, two species belonging to the *Hanseniaspora* genus (*Hanseniaspora osmophila* and *Hanseniaspora valbyensis*), two Candida species (*Candida stellata* and *Candida lactis-condensi*), and *S. cerevisiae* species have been identified (Solieri et al. 2005, 2006). All *Zygosaccharomyces* species recovered from TBV are osmophilic/osmotolerant yeasts growing in media with a high sugar concentration (50 to 60 %), and they are responsible for sugary beverages and food spoilage (Pitt 1975; Fleet 1992; Loureiro and Malfeito-Ferreira 2003). TBV was also the isolation source for a putative new species, provisionally named *Z. pseudorouxii* (Solieri et al. 2007).

Some other TBV-associated species, mainly *Candida* and *Hanseniaspora* spp., are rarely detected in spoiled food and beverages. They are prevalently associated with early stages of wine fermentation, and they can also occur during the middle and late phases. In spite of being a non-osmophilic species, *S. cerevisiae* was found in abundance and it was detected with high frequency in the lowest sugar content cooked musts (Solieri et al. 2006), in agreement with Deak and Beuchat (1996) who found strains of *S. cerevisiae* able to grow in high sugar content foods or beverages. The influence of *S. cerevisiae* and non-*Saccharomyces* yeasts on the flavour of wine and wine vinegar is well understood (Ciani 1998; Fleet 2003). *C. stellata* strains have been found to produce high glycerol, succinic acid, ethyl acetate, and acetoin concentrations that positively influence the aromatic profile of wine vinegar (Ciani 1998). Another two TBV-associated yeasts, *H. osmophila* and *S. ludwigii*, produce high levels of ethyl acetate, acetoin, acetic acid, and acetaldehyde, and they are considered detrimental yeasts in wine fermentation (Ciani and Maccarelli 1998;

Granchi et al. 2002). *S. ludwigii* has been proposed for continuous production of vinegar by Saeki (1990). The role of yeast secondary metabolites in TBV's sensorial features has not yet been studied and a detailed understanding of this aspect is still lacking.

5.3.1 The Management of Vinification

The practice of inoculating cooked must with a selected starter culture or, at least, back-slopping is not widespread. Most fermentation is spontaneous. After cooking, the must is sterile, so the type and number of yeast cells depends on cross-contamination with fresh must, tools, pots, air circulation, and insects: basically a random process that brings some cells to come into contact with the cooked must. The first species able to grow will thus start the fermentation.

Cooked must is a very selective substrate, and only a few genera and species can grow in it because of the low pH, high sugar concentration, and the presence of compounds that inhibit fermentation, such as furfurals. The yeasts involved in the alcoholic fermentation of cooked must belong to different genera and/or species: in spontaneous fermentations the dominance of a strain depends on the specific case, namely the number of cells in the cooked must, the composition of the medium, and environmental conditions like temperature and availability of oxygen.

Fermentation is often driven by different species at different times, due to the continuous modification of the medium favouring one strain at a time. Recently, a starter culture of selected yeast strains was introduced to help control fermentation. Initially, selected strains of *Saccharomyces cerevisiae* for oenological purposes were used, since specific selected strains for TBV were not available. Today, the microbial collection of the University of Modena (www.umcc.unimore.it) has specific selected yeasts for the production of base wine, and the most important traits of the strains suitable for alcoholic fermentation are described in the following paragraph.

The fermentation of cooked must has two key stages, starting and stopping fermentation. In the authors' experience, the final ethanol content of the base wine (BW) should be around 8 % (v/v), sufficient to support acetic fermentation until 6/7 % (w/v) of acetic acid. Once fermentation is started, spontaneously or with a starter culture, it is necessary to stop it promptly as soon as it reaches the desired alcohol content, otherwise the BW will be too high in alcohol and too low in sugar.

The basic reaction of alcoholic fermentation implies a theoretical yield of 51.11 % by weight but, considering the consumption of part of the glucose for plastic purposes and the production of secondary metabolites such as glycerol and succinic acid, the yield is lower. In oenology the average yield is 60 % by volume (46.8 % by weight). In the fermentation of cooked must, the yield is lower (Landi et al. 2005). According to observations, the yield ranges from 35 to 40 % by weight.

Subsequent filtration removes cellular material and other precipitates, which are mainly tartrates whose solubility is modified by the presence of alcohol.

5.3.1.1 How to Prepare a *Pie De Cuve*

The procedure for starting alcoholic fermentation for wine is well established, with selected dried yeasts available on the market. The practice of *pie de cuve* is rarely applied for normal red and white wines, more commonly for champagne, cava, and spumante, but is required for a TBV wine base, since no specific dried yeasts are available on the market.

The first steps to prepare a starter culture are conducted in a lab on a sterile must. A string of cells are transferred from a culture slant to a flask containing 100 ml of sterile must and incubated at room temperature for at least 24 h. The culture is then used to inoculate a volume of 0.5 l of must and then after 24 h this culture is increased to 2.5 l. The production stage starts with 5 l of 24 h culture and the volume is increased every 24 h with filtered cooked must in the ratio of 1:5, one part of culture to five parts of freshly filtered cooked must (Fig. 5.4).

The microbiological quality of the cooked must is very important and the number of yeast cells inside should be as small as possible. The must should never exhibit evident signs of fermentation, with no alcohol and no foam present on the top of the vessel. Filtering the cooked must is not easy and normal wine making equipment is not suitable due to the high viscosity. An old-fashioned vertical cloth filter is the best for homemade TBV (Fig. 5.5), and this kind of filter is also suitable for stopping fermentation.

Fermentation is monitored by alcohol development and immediately stopped as soon as it reaches 8 % v/v of alcohol. The procedure to stop the fermentation involves preliminary filtration to remove the majority of the cells and tartrates, followed by refrigeration or acidification with vinegar. If the filtration is conducted effectively, the BW is microbiologically stable and can be preserved without any other treatment even at room temperature. But the safest and least expensive method is to add vinegar (acidified base wine) until 2.5 % of acetic acid is reached. In this way, the acetic acid content completely inhibits yeast growth and the BW is stable.

5.3.1.2 The Influence of Yeast Species and Strain on the Composition of the Base Wine

The composition of the BW depends significantly on the strain and dominant species of fermenting yeasts. In general, osmophilic species produce less ethanol compared to *S. cerevisiae* and preferentially ferment fructose. They give lower alcoholic yields and produce a greater volume of secondary products. The data available on the influence of yeast strains on the composition of base wine is very limited. They refer to a few important factors for wine making which cannot be

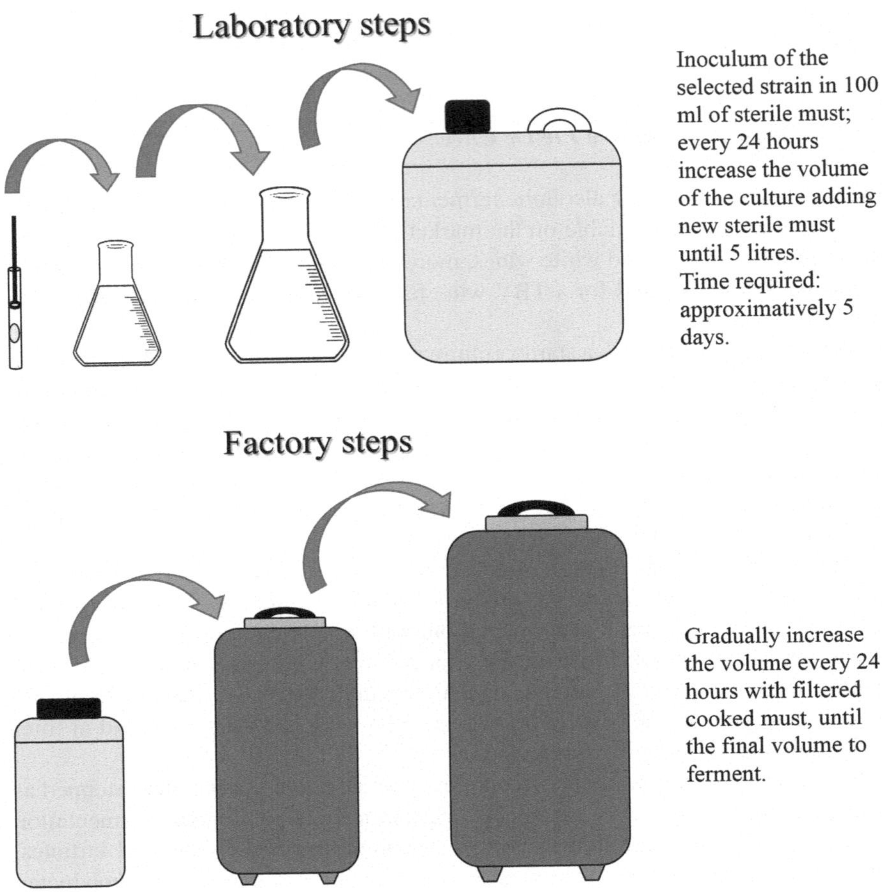

Fig. 5.4 Scaling up from laboratory slant culture to final fermentation vessel

applied to vinegar production because the requirements for a good strain of wine making yeasts are not the same for good vinegar.

The important traits of yeast strains for wine production are well studied and consolidated, with several papers available on the topic. Some phenotypic traits are universal and should always be present in all strains selected for fermentation of food products, in particular for processing non-sterile raw materials. For example, the characteristic of "fitness" is always required at the highest level because this allows the strain to take over from indigenous microorganisms and drive the fermentation (Giudici and Zambonelli 1992; Giudici et al. 2005).

The qualitative characters are more difficult to define and they depend on the product to be obtained. For vinegars, the production of organic acids, such as malic acid and succinic acid, are considered useful, because these compounds influence the fixed acidity, which in turn improves the sensorial properties. In contrast to

Fig. 5.5 Traditional filter cloth to remove lees after alcoholic fermentation

oenological strains, which should not produce hydrogen sulphide because of its unpleasant flavour, for balsamic vinegar this can be positive, because hydrogen sulphide causes precipitation of heavy metals, in particular lead and copper, as insoluble salts that can then be easily eliminated, contributing to the salubrity of the product.

Another very important characteristic is the selective fermentation of glucose, so that the residual sugar is mainly fructose. This offers multiple advantages. Fructose is the most soluble hexose in water and is at least three times as soluble as glucose, in addition to being more hygroscopic and providing greater sweetening power. At the high concentrations of solutes achieved in balsamic vinegars, when the percentage of water can be less than 30 % of weight, the prevalence of fructose over glucose promotes the physical stability of the mixture, limiting the phenomena of sugar crystallisation. Hygroscopicity also increases the retention of water in the system. In the advanced stage of maturation and under specific conditions of high temperature and low relative humidity, this property helps preserve the homogeneity of the mixture. Finally, the sweetening power of fructose (at least twice that of glucose) and its synergy with other substances that make the mixture sweeter and more persistent allow the amount of sugar in balsamic products to be reduced.

In oenological conditions, such as less than 24 % w/w of sugars in the ratio 1:1 glucose fructose, the majority of the yeast species use both the sugars indifferently, and the osmophile yeasts are among the few fructosophile yeast species. The amount of sugar in cooked must is close to 35 % (w/w), a concentration restrictive to the growth of several yeast species and also influencing the metabolism of sugars and secondary fermentation metabolites. From a technological point of view, the osmophile yeast strains, belonging to the *Zygosaccharomyces* genus, and isolated from batches of spontaneously fermented cooked must (Solieri et al. 2006), are the least suitable for cooked must fermentation, since they leave high percentages of unfermented glucose in the media. In contrast, less osmotolerant yeasts, belonging to the genus *Saccharomycodes* and isolated from maturing barrels of TBV, showed a marked preference for glucose (Gobbi et al. 2014).

Two technological traits are mandatory for the selection of yeast strains for TBV production: the ability to ferment cooked must to about 8 % (v/v) of ethanol and the preferential uptake of glucose. For this purpose seven strains belonging to three yeast species were tested in grape must with added sugars up to 40°Bx. Figure 5.6 shows the fermentation curves, which clearly show that strains A6, A7, A8, belonging to *Dekkera bruxellensis* are not osmotolerant and do not grow and ferment in the presence of high concentrations of sugars. The other strains also showed a rather slow growth, but are suitable for the purpose. In particular, the strains B8805 and B8815 (*Saccharomycodes ludwigii*) and the strain A2 (*Hanseniaspora uvarum*) are able to produce alcohol between 6.3 and 7.8 after 32 days.

The data for residual sugars at the end of fermentation are in agreement with previous results and clearly show that the strains tested have a glucose preference (Fig. 5.7). The glucose preference, expressed as the ratio between glucose and fructose consumed, is 1.52, 1.55, and 1.75 for the strains B8805, B8815, and A2, respectively.

The strains belonging to the species *S. ludwigii* show the highest alcohol strength in an osmotic environment and the highest glucose preference. These strains were isolated from TBV and they are the most suitable yeast strains for the fermentation of high sugar and low pH substrates. This makes them ideal candidates for the development of specific starter cultures for fermenting cooked must for balsamic vinegar production.

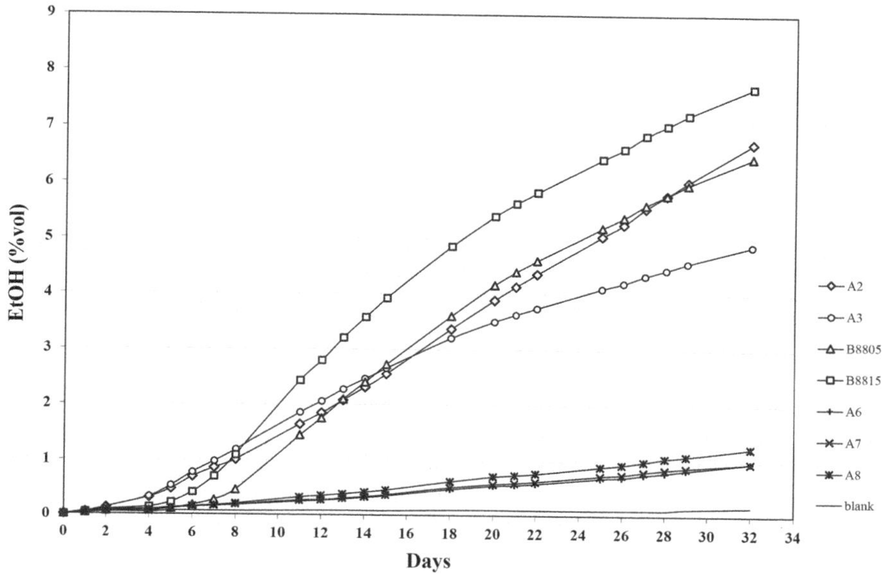

Fig. 5.6 Ethanol produced during the alcoholic fermentation test of grape must enriched with glucose and fructose to simulate the concentration of cooked grape must (410 g/l of sugars, pH 2.8)

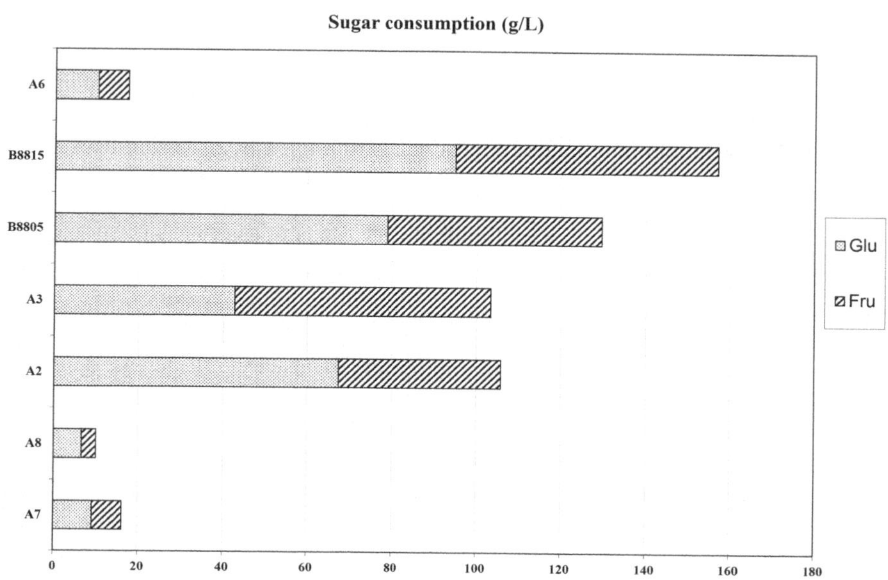

Fig. 5.7 Consumption of glucose and fructose after 32 days of alcoholic fermentation testing of a selective substrate (grape must with 410 g/l of sugars, pH 2.8)

5.4 The Base Vinegar

The production of base vinegar (BVR) is the most difficult step in the whole TBV production process, because it starts with BW, which is a very selective niche medium and involves AAB, which are difficult microorganisms to manage in industrial applications. Acetification should be performed in a different container from the one used for alcoholic fermentation and in static conditions. Traditional static fermentation is chosen for two different reasons: first, the product regulations require that TBV is obtained from traditional fermentation; second, and no less important, to date the attempts to obtain a vinegar base using a submerged culture have proven to be unsatisfactory.

Recent unpublished data from Gullo prove that submerged acetification of BW is possible and advantageous, but requires rigorous control of the fermentation parameters and a two-step fermentation process. In static conditions, the major limit is oxygen transfer and the process usually takes several days or weeks. Other critical factors include temperature control and the overoxidation of ethanol. In fact, AAB are able to complete the oxidation of ethanol into CO_2 and H_2O; this usually happens after exhaustion of ethanol when only acetic acid is present in the medium. To avoid the complete oxidation of ethanol, the process needs to be stopped by filtration to remove the bacterial cells and by stopping oxygen transfer by hermetic closure of the vessel.

It is not easy to decide when to stop fermentation because AAB also oxidise glucose and other molecules with ketonic and aldeidic functions, oxidation products of sensory importance for TBV. Table5.3 reports the data for an extended (94 days) acetic acid fermentation period in wooden barrels in three independent replicas. Ethanol exhaustion occurs after 34 days and starting from that time, a significant amount of gluconic acid is produced from the oxidation of glucose.

The mass balance between the glucose and gluconic acid appears to be incongruent, but this is due to the fact that the barrels were left open and the evaporation of water and other volatile substances causes the concentration of solutes. Consequently the data implies an apparent greater formation of gluconic acid than the amount of oxidised glucose.

Gluconic acid is a relatively strong organic acid and therefore lowers the pH of the vinegar. It is not a volatile substance, and so the fixed acidity is increased. In this way the formation of gluconic acid contributes to maintaining the acidity lost through evaporation of acetic acid during maturation. Gluconic acid can have positive effects on the sweetness of vinegar because of its sensorial properties.

Oxidative fermentation also causes the consumption of glucose with a significant increase in the fructose/glucose ratio. The altered ratio of the two sugars makes products more stable from a physical point of view and more sensorially valued. Finally, gluconic acid was suggested as an indicator of genuineness of balsamic vinegar (Giudici 1993) since its presence, safe cases of fraudulent addition, indicates the direct fermentation of glucose.

Table 5.3 Oxidative over-fermentation of base wine

Time (days)	Ethanol (g/l)	D-glucose (g/l)	D-fructose (g/l)	Acetic acid (g/l)	Acetal-dehyde (mg/l)	Glycerol (g/l)	Gluconic acid (g/l)	L-malic acid (g/l)	D-lactic acid (mg/l)	L-lactic acid (mg/l)
14	11.73	71.33	93.98	35.16	85.59	7.53	8.11	4.28	81.42	47.57
21	8.07	68.85	95.73	43.01	91.77	8.77	9.94	4.18	66.26	136.99
28	3.59	67.00	95.03	47.52	76.58	8.19	13.13	4.19	111.91	116.68
35	1.00	60.62	93.23	51.73	66.24	8.74	15.23	4.07	82.22	136.99
65	0.03	51.11	109.29	53.39	73.63	5.81	38.85	4.69	273.74	230.29
94	0.05	33.44	111.54	34.46	86.62	5.73	62.24	4.71	447.54	353.92

Values are the mean of three batches. A large amount of gluconic acid is produced after the ethanol has been consumed

Acetic acid increases in proportion to the consumption of ethanol, and then decreases at the end of the process, probably due to its complete oxidation into water and carbon dioxide. A significant increase in L- and D-lactic acid and a decrease in glycerol occur upon depletion of ethanol.

In summary, as regards BVR, the most important issue is when to stop the oxidation process. A tentative answer is only possible when the chemical composition of the base vinegar is known, in particular the acidity and glucose content. A simple guideline is that when fixed acidity is low and volatile acidity is high, it is better to leave the AAB to work, while if the opposite is true, it is better to stop acetification. This is because the oxidation of glucose to gluconic acid increases the titratable acidity, and in the meantime the oxidation of acetic acid decreases the pungency of the vinegar (Patent BO2011A000132).

5.4.1 The Scaling Up of Acetification

The oxidative fermentation of the BW usually occurs by spontaneous acetification due to the AAB micro-flora naturally present in the environment. Recently, the use of selected AAB strains has been proposed, with the development of a procedure for scaling up the fermentation process to vinegar factory scale (Gullo and Giudici 2008). The starting point is a freshly prepared culture of a selected strain, which is monitored to assess bacterial growth. After about a fortnight, the volume of culture is sufficient for transfer to the factory for the inoculation of a larger volume of BVR and then finally for inoculating the static fermentation vat (Fig. 5.8).

Two simple empirical rules should be followed to ensure success:

1. To increase the volume of the culture, always put the new BW into the culture vessel, never vice versa;
2. The amount of BW for each top up is calculated by the 2.5 % rule. This means that the alcohol and acetic acid concentration in the culture should never go below 2.5 % (v/v for alcohol, w/v for acetic acid), and the amount of BW to add to the culture should be calculated considering the volume, the acidity and the alcohol content of the culture (Gullo et al. 2009).

5.4.2 The Selection of Acetic Acid Bacteria

Recent studies on AAB of TBV reported the presence of species already identified in other vinegars, including *Gluconacetobacter europaeus*, *Gluconacetobacter hansenii*, *Gluconacetobacter xylinus*, *Acetobacter pasteurianus*, *Acetobacter aceti*, and *Acetobacter malorum* (De Vero et al. 2006; Gullo et al. 2006; Gullo and Giudici 2008). Among the species identified to date, strains of *Ga. europaeus* appear to be the most widespread. AAB are Gram-negative, strictly aerobic, and by

Fig. 5.8 Scaling-up protocol for acetification of base vinegar

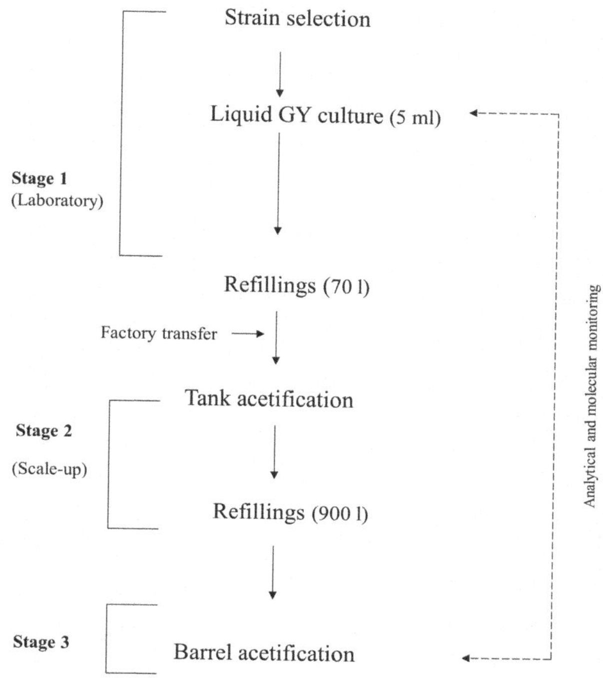

respiratory metabolism they are able to oxidise a wide range of organic compounds with hydroxyl, aldehyde, and ketone groups, such as alcohols, polyols, and sugars.

In general, all products of alcoholic fermentation can be oxidised by acetic acid bacteria: ethanol with the formation of acetic acid, other alcohols like *n*-propanol, isobutanol, and isoamyl alcohol into the corresponding organic acids, glycerol to dihydroxyacetone, and glucose to gluconic acid. Acetic acid bacteria preferentially oxidise ethanol and only in its absence do they convert other alcohols, since the presence of ethanol appears to inhibit the use of other forms of energy (Asai 1968), with a significant impact on the composition of the final product.

Acetic acid fermentation, both industrial and small scale, is always achieved with mixed cultures of acetic acid bacteria from previous batches or with indigenous bacteria present in the raw materials. The use of selected strain cultures is only an interesting research topic that is implemented because vinegar is generally a cheap commodity and also because the selective pressure changes during the fermentation process with a succession of species appearing.

However, the high phenotypic and genetic variability exhibited by AAB is a significant feature suggesting the possibility of designing genetic improvements of strains for industrial applications and, in this specific case, for balsamic vinegar production. Phenotypic variability is very high in both intra- and inter-species, with some species more alcohol tolerant and others more acidophilic. Only three genera include important species for vinegar: *Acetobacter*, *Gluconobacter*, and

Gluconacetobacter. In general, strains belonging to the species *Acetobacter pasteurianus* and *Acetobacter aceti* are very alcohol tolerant and acidophilic, which makes them suitable for use in industrial fermenters. Tolerance to acetic acid is a quantitative trait involving many genes and is also influenced by the composition of the medium. For example, strains able to produce exopolysaccharides, if grown in a static system, greatly increase their tolerance to acetic acid thanks to the formation of thick films.

All selected strains must have some universal characteristics, while others are more specific to the technology used and the product to be obtained. For the acidification of the BW, the strains should be tolerant to about 25 % by weight of sugars, while resistance to ethanol is less important, since the alcohol content of BW is rarely above 8 % (v/v). Other important selection characters are the ability to preferentially consume ethanol with rapid formation of acetic acid to prevent the growth of other microorganisms, the resistance to bacteriophage infections, non-production of negative sensorial molecules, and the ability to grow in laboratory media (Gullo and Giudici 2008).

The conversion yield of ethanol is not so important for the production of TBV, the key traits instead being the ability to form films with a low content of exopolysaccharides, the ability to grow at low pH values, the tolerance to high concentrations of sugars, and the ability to produce significant amounts of gluconic acid. On this basis, a project was started for the selection of AAB strains suitable for fermenting the base wine, considering those belonging to the two dominant TBV species: *A. pasteurianus* and *G. europaeus* (Gullo et al. 2009).

5.5 The Maturing Process

The barrel set procedure, maturation, and yield are described in Chap. 6. The authors have introduced two significant improvements to the traditional way of managing the barrel set:

- The biological process is completely separated from maturation, the latter occurring in the barrel set where only chemical and physical modifications occur;
- The barrels are completely filled with base vinegar (BVR) and sealed.

Traditionally, the barrels were not completely filled and the holes on the top were covered with river stones, while today they are covered with cloths (Fig. 5.9).

The decision to completely fill the barrels and seal them was based on the need to keep as much as possible of the volatile and sensorially active compounds of the TBV inside the barrel. These potentially volatile molecules with low vapour pressure are lost through the open barrel holes together with water, and only solids are retained inside the barrel.

In contrast, completely filled and sealed barrels do not freely lose volatile molecules, and the mass transfer of water and other molecules is mediated by the

Fig. 5.9 Barrel sets for the ageing of Traditional Balsamic Vinegar. Top holes are left open and covered with cloths

wood of the barrels. The wooden casks act as molecular filter: low steric dimension molecules pass selectively throughout the wood pores while larger molecules are retained (Siau 1984; Lemmetti and Giudici 2010) resulting in a general increase in solute inside the vinegar bulk.

Table 5.4 reports the steric size of some molecules characteristic of TBV and potentially involved in the concentration process, calculated using the method of Connolly. This method simulates the solvent as a sphere and maps the points of contact with the Van der Waals atomic radius of the molecule. The set of these points constitutes the surface accessible to solvents, and the volume enclosed within is the *solvent excluded volume*, which very closely approximates the volume of the molecule. The simulation was performed in the minimum energy conformation, i.e. with the spatial arrangement of atoms such that the overlapping of atomic orbitals involved in the bonds are the closest possible to the theoretical value that determines the highest energy, but simultaneously minimises the electronic repulsions between non-bonded atoms.

The selective transfer of molecules through the wood can be explained on the basis of their steric hindrance or molecular size. The vapour pressure, related to the tendency to evaporate, is not related to size and molecular weight. It depends on the intermolecular Van der Waals forces and hydrogen bonds responsible for the condensed states of the matter. For example, the ester ethyl acetate has a high

Table 5.4 Relevant properties of some compounds involved in the volatilisation process during the maturing of TBV in wooden barrels

Compound	Molecular weight (g/mol)	Relative density $d_{20/20}$	Vapor pressure at 25°C (KPa)	Connolly surface (Å^2)	Connolly molecular volume (Å^3)
Water	18.0	1.00	3.17	23.3	10.3
Acetic acid	60.0	1.04	2.07	73.1	46.9
Butanoic acid	88.1	0.95	0.221	111.9	83.2
Ethanol	46.1	0.79	5.73	74.1	48.6
2-Methyl-1-propanol	74.1	0.80	2.39	109.2	84.3
2,3-Butandiol	90.1	1.00	0.02	114.3	90.8
Acetaldehyde	44.0	0.78	120	65.4	40.7
Furfural	96.1	1.16	0.29	100.1	70.3
Ethylacetate	88.1	0.90	12.6	114.6	81.8
Ethylbutanoate	116.2	0.87	2.01	152.4	116.5

Relative density and vapour pressure are from the CRC Handbook of Chemistry and Physics, 88th Ed.; surface and volume of molecules are calculated in the minimum energy conformation (determined with the MM2 algorithm) with the Connolly tool in ChemBio3D Ultra v.11

volatility, but its volume is eight times greater than that of water, making it difficult to pass through the pores of the wood.

In short, the barrel set behaves essentially as a device for vinegar concentration. Water passes through the wood of the barrel and evaporates into the environment. The loss of water and other volatile substances through the hole on the top of the cask is different from the loss through the barrel wood. The fraction evaporated from the top is mainly due to vapour pressure, with a greater molar fraction of volatile compounds. Furthermore, hermetically closed barrels improve TBV quality because it helps to preserve the vinegar's aromatic compounds, which often have high molecular weights and volumes, but low vapour pressures.

The transfer of the liquid from the inside to the outer surface of the barrel occurs in two different phenomena topologically separated. First, the liquid impregnates the wood and the front of impregnation in some condition can be extended enough to reach the outer surface. Secondly, in the non-impregnated wood, the transfer of molecules occurs with change of state, from liquid to gas. In general, the higher the relative humidity of the surroundings, the greater the impregnation layer that will eventually reach the outer surface of the barrel. Figure 5.10 shows a clear example of how the front of impregnation has reached the outer surface of the barrel and, due to the humidity of the wood surface, fungal mycelium has grown.

The rate of water transfer through the wooden casks is influenced by several factors, including the hygroscopicity of the vinegar (or water activity), the relative humidity of the surrounding air, the type of wood, and the thickness, size, and shape of the cask. Furthermore, the mass transfer of vinegar throughout wooden barrels involves the following factors:

Fig. 5.10 Top sealed barrel set for ageing Traditional Balsamic Vinegar. Abundant white mycelium is visible on the surface of the barrels

- A transition time is required to reach the outside of the barrel.
- Evaporation losses consist mainly of water.
- The mass flux throughout the barrel is a function of wood type.
- The mass flux is more pronounced when the vinegar is in contact with the wood.
- The mass flux depends on ambient conditions.

The most widely used model describing the phenomenology of mass transfer throughout wooden barrels assumes a series of three mechanisms, i.e. absorption, diffusion, and evaporation. Mass transfer within the free water domain is driven by pressure potential according to Darcy's law. Depending on the porosity of the wood, the impregnation front can take more or less time to reach the outside surface. Fick's law governs the flow of vinegar within the hygroscopic domain by the diffusion mechanism. The lower the relative humidity of the air, the higher the water transfer. However, the mass diffusion throughout the wood represents the rate-limiting stage of the vinegar transfer throughout a barrel, and Fick's second law quantitatively describes how diffusion affects the concentration profile under isothermal conditions, directly proportional to the diffusion coefficient.

In a simplified model, diffusion occurs under a moisture gradient in wood and the mass flux is proportional to how far the wood water content is from its saturation equilibrium point. This means that vinegar evaporates towards the outside of the barrel if the wood water content is lower than the outside ambient one. Mass transfer occurs until the driving forces can overcome the opposing resistances. In addition to the relative humidity of the ambient, other factors affect the evaporation rate from the external surface of the barrels, such as temperature and air velocity. Vinegar maturing causes water activity to decrease due to the increase of solute interaction, and so the water activity of vinegar will become the limiting driving factor for evaporation. In general, lower relative humidity makes vinegar concentration faster.

The diffusion of molecules through the wood is a well-known phenomenon in the maturation process of wine and brandy, occurring in sealed barrels. For wines, the volume loss per year ranges from 1 to 9 % (Canas et al. 2002). In the southern hemisphere, in areas with low relative humidity, product losses can reach 15 % in 2 years of maturing in barrels. In the world of whiskey, the product loss is known as the "angel's share".

The common practice of keeping the barrels not completely filled and with a hole on the top covered by cloth has several disadvantages. First, the head space increases the evaporating surface but decreases the contact of the vinegar with the wood and consequently reduce its selective capacity. Second, the large hole promotes the release of sensorially active molecules with low vapour pressure, and a large amount of aromatic substances are lost. Third, the loss of volatile compounds, other than water, decreases the yield and the residence time of the TBV. Finally, the open holes are a potential pathway for insects reaching the vinegar. The authors strongly discourage this practice for all of these reasons.

References

Amerine MA, Kunkee RE (1968) Microbiology of winemaking. Annu Rev Microbiol 22:323–358

Asai T (1968) Oxidation of other aliphatic monoalcohols. In: Acetic acid bacteria: classification and biochemical activities. University of Tokyo Press, Japan, pp 141–151

Benedetti B (2004) Fare in casa l'aceto balsamico, 2nd edn. Edizioni Il fiorino, Modena

Canas S, Belchior P, Mateus AM, Spranger MI, Bruno-de-Sousa R (2002) Kinetics of impregnation/evaporation and release of phenolic compounds from wood to brandy in experimental model. Ciência Téc Vitic 17:1–14

Ciani M (1998) Wine vinegar production using base wines made with different yeast species. J Sci Food Agric 78:290–294

Ciani M, Maccarelli F (1998) Oenological properties of non-Saccharomyces yeasts associated with wine-making. Word J Microbiol Biotechnol 14:199–203

Davenport RR (1974) Microecology of yeast and yeast-like organisms associated with and English vineyard. Vitis 13:123–130

Deak T, Beuchat LR (1996) Handbook of food spoilage yeasts. CRC, Boca Raton, FL

De Vero l, Gala E, Gullo M, Solieri L, Landi S, Giudici P (2006) Application of denaturing gradient gel electrophoresis (DGGE) analysis to evaluate acetic acid bacteria in traditional balsamic vinegar. Food Mirobiology 23:809–813

Falcone PM, Giudici P (2010) Sugar conversion induced by the application of heat to the grape must. J Agric Food Sci 58:8680–8691

Fleet GH (1992) Spoilage yeast. Crit Rev Microbiol 12:1–44

Fleet GH (2003) Yeast interactions and wine flavour. Int J Food Microbiol 86:11–22

Fleet GH, Heard GM (1993) Yeasts-growth during fermentation. In: Fleet GH (ed) Wine microbiology and biotechnology. Harwood, Chur, Switzerland, pp 27–55

Gobbi M, De Vero L, Solieri L, Comitini F (2014) Fermentative aptitude of non-Saccharomyces wine yeast for reduction in the ethanol content in wine. European Food Research and Technology 239:41–48

Giudici P (1990) Inibizione dell'acido acetico sull'attività dei lieviti osmofili isolati dall'aceto balsamico tradizionale. Industrie delle Bevande 110:475–480

Giudici P (1993) Acido gluconico: criterio di genuinità nell'aceto balsamico tradizionale. Industrie delle Bevande 124:123–125

Giudici P, Zambonelli C (1992) Criteri di selezione dei lieviti per enologia. Vignevini 9:29–34
Giudici P, Altieri C, Cavalli R (1992) Aceto balsamico tradizionale, preparazione del fermentato di base. Industrie delle Bevande 122:478–483
Giudici P, Solieri L, Pulvirenti AM, Cassanelli S (2005) Strategies and perspectives for genetic improvement of wine yeasts. Appl Microbiol Biotechnol 66:622–628
Granchi L, Ganucci D, Messini A, Vincenzini M (2002) Oenological properties of Hanseniaspora osmophila and Kloeckera corticis from wines produced by spontaneous fermentations of normal and dried grapes. FEMS Yeast Res 2:403–407
Gullo M, Giudici P (2008) Acetic acid bacteria in traditional balsamic vinegar: phenotypic traits relevant for starter cultures selection. Int J Food Microbiol 125:46–53
Gullo M, Caggia C, De Vero L, Giudici P (2006) Characterization of acetic acid bacteria in traditional balsamic vinegar. Int J Food Microbiol 106:209–212
Gullo M, De Vero L, Giudici P (2009) Succession of selected strains of Acetobacter pasteurianus and other acetic acid bacteria in Traditional Balsamic Vinegar. Appl Environ Microbiol 75 (8):2585–2589
Landi S, Castellari L, Gullo M, Solieri L, De Vero L, Giudici P (2005) Yeasts and sugar of traditional balsamic vinegar. In: Vinegars and acetic acid bacteria international symposium, Reggio Emilia, Italy
Lemmetti F, Giudici P (2010) Bilancio di massa ed età dell'Aceto Balsamico Tradizionale. Industrie delle Bevande 230:18–28
Loureiro V, Malfeito-Ferreira M (2003) Spoilage yeasts in wine industries. Int J Food Microbiol 86:23–50
Muratore G, Licciardello F, Restuccia C, Puglisi ML, Giudici P (2006) Role of different factors affecting the formation of 5-hydroxymethyl-2-furancarboxaldehyde in heated grape must. J Agric Food Chem 54:860–863
Patent No BO2011A000132 (Italy) Procedimento per la produzione di un condimento agrodolce invecchiato di mosto d'uva cotto
Pitt JI (1975) Xerophilic fungi and the spoilage of foods of plant origin. In: Duckworth RB (ed) Water relations of foods. Academic, London, pp 273–307
Sacchetti M (1974) Sull'aceto balsamico modenese. Edagricole, Bologna
Saeki A (1990) Studies on acetic acid fermentation. III continuous production of vinegar with immobilized Saccharomycodes ludwigii cells and immobilized Acetobacter aceti cells entrapped in calcium alginate gel beads. J Jpn Soc Food Sci 37:722–725
Siau JE (1984) Transport process in wood. Springer, Berlin
Solieri L, Giudici P (2008) Yeasts associated to Traditional Balsamic Vinegar: ecological and technological features. Int J Food Microbiol 125(1):36–45
Solieri L, Gullo M, De Vero L, Antunez O, Perez-ortin JE, Giudici P (2005) Homogeneity of interspecific hybrids between *Saccharomyces cerevisiae* and *Saccharomyces uvarum* by phenotypic and transcriptional analysis. Int J Biotechnol Biochem 1:11–21
Solieri L, Landi S, De Vero L, Giudici P (2006) Molecular assessment of indigenous yeast population from traditional balsamic vinegar. J Appl Microbiol 101(1):63–71
Solieri L, Cassanelli S, Giudici P (2007) A new putative Zygosaccharomyces yeast species isolated from traditional balsamic vinegar. Yeast 24:403–417
Turtura GC, Benfenati L (1988) Caratteristiche microbiologiche e chimiche dell'Aceto Balsamico Naturale. Studio del prodotto. Ann Microbiol 38:51–74

Chapter 6
The Ageing of Balsamics: Residence Time, Maturity, and Yield

It is a common opinion that *matured*, referring to balsamics, means selective products of high quality and special sensorial properties. This general idea is supported by advertisements, marketing strategies, gourmets, chefs, and dealers and is widespread on Internet. It is obvious that balsamic age is considered a very distinctive trait, and many balsamic products claim or fake the improbable age of 100 years. This can easily be verified by launching an Internet search with key-words like “balsamic vinegar” and “age”. However, before going into the details and discussing the honesty of these claimed vinegar ages, it is worth defining the important issues of what age is, residence time, and yield and the relationship between these important variables.

In the hypothesis of a single barrel, filled with freshly made vinegar 20 years ago, the question as to how old the vinegar inside the barrel is now seems obvious: 20 years. More precisely, 20 years is the time that vinegar has been inside the barrel, and this is defined as *residence time* (RT). However, the *age* of the vinegar is best defined by the degree of chemical modifications of its constituents. In brief, the physical dimension of RT is time, while age is the physical and/or chemical modification of the vinegar constituents and could be better defined as “maturity”.

To reinforce this concept, consider a vinegar divided into two lots and aged in two different environments for the same RT. This results in two lots of vinegar of different chemical composition. One is more mature than the other. This supposition does not require further proof, since it is general knowledge that the shelf life of foods depends on the way they are stored. The same applies to balsamic vinegar, which during storage in a barrel set undergoes physical and chemical modifications considered to be positive because they increase the quality of the vinegar. RT does not require further explanation, it is the interval of time between the filling of the barrel with vinegar and its removal for bottling and sale.

In general, the calculation of RT is easy but for TBV is a little more complicated because of the peculiar management of the barrel set. The most appropriate definition of "maturity" of balsamic product is possibly the following: “the degree

P. Giudici et al., *Balsamic Vinegars*, DOI 10.1007/978-3-319-13758-2_6

of positive modification that occurred during the RT", including as it does a quantitative indication (degree) of a qualitative parameter (positive modification).

The point is to establish the chemical and physical variables correlated with the quality of balsamics and related to the RT. Another very important parameter, because it has an appreciable impact on the cost of production, is *yield*. This is the ratio between the amount of finished balsamic product and the initial amount of concentrated and cooked must (expressed either in volume or in weight). For TBV, yield and RT are two directly related parameters, and the higher the balsamic yield, the higher the RT and, as a consequence, also the maturity of the vinegar. In the following paragraphs all these relationships are detailed and supported with data and mathematical models. For the other balsamics, ageing is not significant, with the single exception of BVM which is matured for an RT of at least 60 days, with some producers sometimes claiming longer periods, but never as long as for TBV.

The first aspect of ageing to be considered is the RT, a variable easily determined for BVM because it is completed in a single vessel, filled and emptied completely for each lot of vinegar. It is more complicated to calculate RT for TBV because of the set of barrels and the transfer of portions of vinegar between the barrels of the set. Every year only a small amount vinegar is removed from the smallest barrel, which is then refilled with vinegar from the barrel of next size up and so on for each barrel of the set. The largest barrel is topped up with new base vinegar (Fig. 6.1).

TBV is tapped off from the smallest barrel. Then each barrel is refilled with the vinegar from the barrel on the left and the largest is refilled with new BVR. Only a fraction of the volume of vinegar is transferred from barrel to barrel. The result is a blending of vinegar of different ages in each barrel. This procedure is called *rincalzo* ("refilling" or "topping up" are acceptable translations), and the purpose is to replace the vinegar tapped off and/or lost by evaporation and/or transferred from barrel to barrel. Because of this procedure, each barrel contains a blend of vinegars of different ages and obviously the age increases from the largest to the

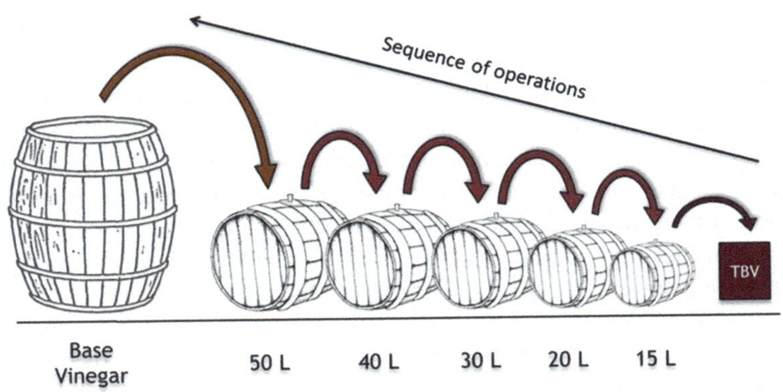

Fig. 6.1 Schematic representation of a typical barrel set for traditional balsamic vinegar (TBV) maturing. The number and capacity of the barrels can vary

smallest barrel, and it is also influenced by the volumes of the barrels and by the volume of vinegar transferred from barrel to barrel (see Chap. 4).

This refilling procedure produces a blend of vinegars of different RT in each barrel. A mathematical model was developed to calculate the RT of the vinegar in each barrel and of the TBV tapped off (Giudici and Rinaldi 2007).

6.1 Mathematical Models to Calculate the Residence Time of Traditional Balsamic Vinegar

The theoretical model developed for calculation of RT is based on the solution of the Zeno's paradox of Achilles and the tortoise. In detail, the RT (in agreement with the symbols used in the original paper, E_i is used as an abbreviation of RT of the barrel i in the following equations) is a sequence of real numbers, and the mathematical study of these sequences tells us the real E_i of vinegar in each barrel at year n. The theoretical study of these sequences shows that there is an upper bound for the RT of vinegar, and this upper bound can be formulated through the values of the volumes of vinegar in the barrel and transferred, respectively, V_i and R_i.

It proves that the sequences E_i are strictly increasing and with a finite limit as n increases towards infinity, and these limits are calculated. It is relatively simple to implement a computer procedure (attached to this book) to calculate each $E_i(n)$ at a fixed n.

To establish a simple notation, a set of five barrels is considered, even though the model works with a set of t barrels, for each $t \geq 1$. The barrels are numbered from the smallest to the largest with the integer numbers from 1 to 5. V_i, $1 \leq i \leq 5$ denotes the volume of vinegar contained in the i-th barrel at the end of the annual procedure of refilling. Obviously, V_i is the same every year. It is supposed that the volume of the removed TBV from barrel 1, denoted by V_p, is the same every year as well as the volume lost by evaporation in each barrel. P_i denotes the volume that is lost every year by evaporation in barrel number i. Under these hypothesis, the volume R_i transferred from the $(i+1)$-th barrel to the i-th barrel each year is constant, and it is obtained by summing up V_p with all the P_j, $j \leq i$. Since the procedure of refilling happens once a year, the years are numbered with the set N of natural numbers, and for each $n \in N$, $E_i(n)$ denotes the age of vinegar contained in the i-th barrel at year n and at the end of the annual procedure of refilling. The year 0 corresponds to the beginning of the process, when all barrels are filled with new base vinegar. Therefore, $E_i(0) = 0$ is set for each i.

To define $E_i(n)$, it is observed that the annual procedure is conducted in the following order:

- V_p is removed from barrel number 1.
- R_i is transferred from barrel $i+1$ to barrel i starting from barrel 2.

Therefore, the age of V_p at year n is $E_1(n-1)+1$.

The age of R_i at year n is $E_{i+1}(n-1)+1$ for each $1 \leq i \leq 4$. The age of R_5 is the age of the must, and this age is set to 0. E_i is defined as the pondered mean between the age of vinegar in the i-th barrel before refilling and the age of R_i at year n.

More precisely:

$$E_5(n) = \frac{R_5 \cdot 0 + (V_5 - R_5) \cdot (E_5(n-1)+1)}{V_5}$$
$$= \frac{(V_5 - R_5) \cdot E_5(n-1)}{V_5} + \frac{V_5 - R_5}{V_5}. \tag{6.1}$$

$$E_4(n) = \frac{R_4 \cdot (E_5(n-1)+1) + (V_4 - R_4) \cdot (E_4(n-1)+1)}{V_4}$$
$$= \frac{R_4}{V_4} E_5(n-1) + \frac{V_4 - R_4}{V_4} E_4(n-1) + 1. \tag{6.2}$$

$$E_3(n) = \frac{R_3 \cdot (E_4(n-1)+1) + (V_3 - R_3) \cdot (E_3(n-1)+1)}{V_3}$$
$$= \frac{R_3}{V_3} E_4(n-1) + \frac{V_3 - R_3}{V_3} E_3(n-1) + 1. \tag{6.3}$$

$$E_2(n) = \frac{R_2 \cdot (E_3(n-1)+1) + (V_2 - R_2) \cdot (E_2(n-1)+1)}{V_2}$$
$$= \frac{R_2}{V_2} E_3(n-1) + \frac{V_2 - R_2}{V_2} E_2(n-1) + 1. \tag{6.4}$$

$$E_1(n) = \frac{R_1 \cdot (E_2(n-1)+1) + (V_1 - R_1) \cdot (E_1(n-1)+1)}{V_1}$$
$$= \frac{R_1}{V_1} E_2(n-1) + \frac{V_1 - R_1}{V_1} E_1(n-1) + 1. \tag{6.5}$$

So what is the real E_i of vinegar in each barrel?

First of all a preliminary property is identified whose proof is straightforward.

Property 1
The following equality holds for each real number x and for each positive integer n:

$$\sum_{i=1}^{n} x^i = \frac{x^{n+1} - x}{x - 1}. \tag{6.6}$$

Looking at E_5, we have:

$$E_5(1) = \frac{V_5 - R_5}{V_5}.$$

$$E_5(2) = \frac{(V_5 - R_5)\cdot(E_5(1)+1)}{V_5} = \frac{V_5 - R_5}{V_5} + \frac{(V_5 - R_5)^2}{V_5{}^2}.$$

$$E_5(3) = \frac{(V_5 - R_5)\cdot(E_5(2)+1)}{V_5} = \frac{V_5 - R_5}{V_5} + \frac{(V_5 - R_5)^2}{V_5^2} + \frac{(V_5 - R_5)^3}{V_5^3}.$$

$$E_5(n) = \sum_{i=1}^{n} \frac{(V_5 - R_5)^i}{V_5^i}. \tag{6.7}$$

Obviously, E_5 is increasing.

Set $x = (V_5 - R_5)/V_5$ and apply Property 1:

$$E_5(n) = \frac{\left(\frac{V_5-R_5}{V_5}\right)^{n+1} - \frac{V_5-R_5}{V_5}}{\frac{V_5-R_5}{V_5} - 1} = \frac{\frac{V_5-R_5}{V_5}\left[\left(\frac{V_5-R_5}{V_5}\right)^n - 1\right]}{-\frac{R_5}{V_5}}$$
$$= \frac{V_5 - R_5}{R_5}\left[1 - \left(\frac{V_5 - R_5}{V_5}\right)^n\right]. \tag{6.8}$$

Since $0 < ((V_5 - R_5)/V_5) < 1$, it is concluded that:

$$\lim_{n\to+\infty} E_5(n) = \frac{V_5}{R_5} - 1. \tag{6.9}$$

This is the upper bound for the age of vinegar in barrel number 5.

It now proves that each sequence E_i, $1 \le i \le 4$, is strictly increasing and upper bounded as well as E_5. This establishes that each sequence E_i has a finite limit which can then be calculated.

It is already known that E_5 increases; therefore, it is sufficient to prove that E_{i-1} increases whenever E_i does. Proceeding by induction on n:

Suppose $E_{i-1}(n) > E_{i-1}(n-1)$, then:

$$E_{i-1}(n+1) = \frac{R_{i-1}}{V_{i-1}}E_i(n) + \frac{V_{i-1} - R_{i-1}}{V_{i-1}}E_{i-1}(n) + 1$$
$$> \frac{R_{i-1}}{V_{i-1}}E_i(n-1) + \frac{V_{i-1} - R_{i-1}}{V_{i-1}}E_{i-1}(n-1) + 1 = E_{i-1}(n)$$

and the assertion follows.

Now it is proven that E_i is upper bounded for each i, $1 \le i \le 5$.

It is already known that E_5 is upper bounded; therefore, it is sufficient to prove that E_{i-1} is upper bounded whenever E_i is upper bounded.

Let $E_i(n) \le M$, for each $n \in N$, with M a fixed positive real number. Giving:

$$E_{i-1}(n) = \frac{R_{i-1}}{V_{i-1}} E_i(n-1) + \frac{V_{i-1} - R_{i-1}}{V_{i-1}} E_{i-1}(n-1) + 1. \tag{6.10}$$

Set $R_{i-1}/V_{i-1} = s$ and $(V_{i-1} - R_{i-1})/V_{i-1} = t$, therefore:

$$\begin{aligned} E_{i-1}(1) &= 1; \\ E_{i-1}(2) &= sE_i(1) + t + 1; \\ E_{i-1}(3) &= sE_i(2) + stE_i(1) + t^2 + t + 1; \\ E_{i-1}(4) &= sE_i(3) + stE_i(2) + st^2E_i(1) + t^3 + t^2 + t + 1. \end{aligned}$$

In general:

$$E_{i-1}(n) = \sum_{j=0}^{n-2} st^j E_i(n-j-1) + \sum_{j=0}^{n-1} t^j \le Ms \sum_{j=0}^{n-2} t^j + \sum_{j=0}^{n-1} t^j; \tag{6.11}$$

observe that $0 < t < 1$ and applying Property 1:

$$\sum_{j=0}^{n-2} t^j = 1 + \frac{t^{n-1} - t}{t-1} < 1 + \frac{t}{1-t} \quad \text{and} \quad \sum_{j=0}^{n-1} t^j = 1 + \frac{t^n - t}{t-1} < 1 + \frac{t}{1-t},$$

gives:

$$E_{i-1}(n) < Ms + \frac{Mst}{1-t} + 1 + \frac{t}{1-t} \tag{6.12}$$

and the assertion follows.

It is now possible to calculate the limits. First the following Property 2 is recalled (its proof is straightforward).

Property 2
The following equality holds for each i, $1 \le i \le 5$.

$$\lim_{n \to +\infty} E_i(n) = \lim_{n \to +\infty} E_i(n-1). \tag{6.13}$$

Set $l_i = \lim_{n \to +\infty} E_i(n)$.
It is already known that $l_5 = ((V_5/R_5) - 1)$, let $i \ne 5$ and calculate:

$$\begin{aligned} \lim_{n \to +\infty} E_i(n) &= \lim_{n \to +\infty} \frac{R_i \cdot (E_{i+1}(n-1) + 1) + (V_i - R_i) \cdot (E_i(n-1) + 1)}{V_i} \\ &= 1 + \frac{R_i}{V_i} \lim_{n \to +\infty} E_{i+1}(n-1) + \frac{V_i - R_i}{V_i} \lim_{n \to +\infty} E_i(n-1). \end{aligned}$$

From Property 2:

$$\left(1 - \frac{V_i - R_i}{V_i}\right) \lim_{n\to+\infty} E_i(n) = 1 + \frac{R_i}{V_i} \lim_{n\to+\infty} E_{i+1}(n),$$

giving:

$$\lim_{n\to+\infty} E_i(n) = l_{i+1} + \frac{V_i}{R_i}. \tag{6.14}$$

Concluding with the following result:

E_5 limit in barrel number 5: $\frac{V_5}{R_5} - 1$
E_4 limit in barrel number 4: $\frac{V_5}{R_5} + \frac{V_4}{R_4} - 1$
E_3 limit in barrel number 3: $\frac{V_5}{R_5} + \frac{V_4}{R_4} + \frac{V_3}{R_3} - 1$
E_2 limit in barrel number 2: $\frac{V_5}{R_5} + \frac{V_4}{R_4} + \frac{V_3}{R_3} + \frac{V_2}{R_2} - 1$
E_1 limit in barrel number 1: $\frac{V_5}{R_5} + \frac{V_4}{R_4} + \frac{V_3}{R_3} + \frac{V_2}{R_2} + \frac{V_1}{R_1} - 1$

Since the E_i of the removed vinegar at year n is $E_1(n-1)+1$, the limit of the age of the removed vinegar is:

$$\frac{V_5}{R_5} + \frac{V_4}{R_4} + \frac{V_3}{R_3} + \frac{V_2}{R_2} + \frac{V_1}{R_1}.$$

The model can be repeated if a set of t barrels is used with t an integer greater or equal to 1. Using similar notations, it is concluded in the same manner that for each i, with $1 \le i \le t$, the age limit in barrel number i is $\sum_{j=i}^{t} (V_j/R_j) - 1$.

Moreover, the age limit of the removed vinegar is $\sum_{j=1}^{t} (V_j/R_j)$.

In reality, in a real barrel set, the tapping off of TBV is not the only output from the barrel set. There are structural losses (the barrels may leak), and water is lost from the system through the walls of the barrels as evaporation. This means that the yield (the yield, on a volume basis, is the percentage of the volume of TBV tapped off divided by the volume of refilled vinegar) is strictly dependent to the amount of leakage and evaporation of volatiles. Consequently, to calculate the yield of a barrel set, it is necessary to consider all the losses, both structural and evaporated volatiles (Fig. 6.2).

These data are difficult to evaluate directly, but they are easily derived from the difference between the annual volume of base vinegar used for refilling the barrel set and the volume of the vinegar removed.

A basic computer program, based on the model above, was developed using an Excel spreadsheet and is available with this book. A hypothetical barrel set of five elements was considered, all completely filled and with decreasing volumes, respectively, of 50, 40, 30, 20, and 15 l, for a total volume of 155 l. The amount of annual refilling, yield, and TBV removed was simulated to vary in ranges according to the averages observed for real barrel sets. Specifically, the volume

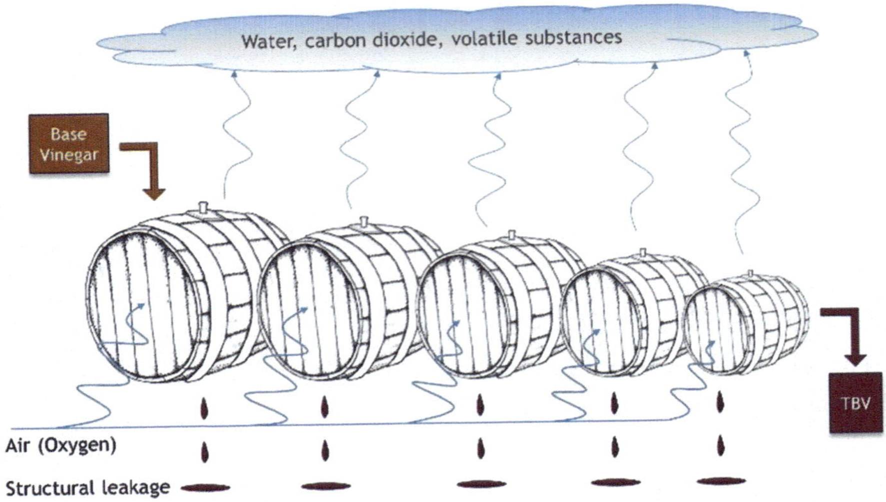

Fig. 6.2 Representation of mass fluxes of a barrel set for ageing of traditional balsamic vinegar

of TBV collected every year varies between 0 and 6 l (6 l is 40 % of the volume of the smallest barrel or about 3.9 % of the total volume of the barrel set); the volume of base vinegar refilled is 0–50 l (50 l is the capacity of the largest barrel; the yield does not exceed 40 %. These limits impose maximum losses (evaporation) of about 31 %.

Figure 6.3 shows the curve obtained from the simulation of yield as a function of the refilling level and at different levels of TBV removal. The yield is strictly dependent on the refilling required for the TBV removed; in other words, the yield is higher when the losses are low.

To illustrate the direct influence of losses, mainly represented by evaporation, on the amount of TBV removed, the evaporation level was varied, expressed as a % of the annual volume lost from the barrel set, at two constant levels of RT: 12 and 25 years (Fig. 6.4). It is immediately obvious that for constant RT, the amount of TBV removed is inversely proportional to the evaporation level.

RT is high when the sum of the amount of TBV removed plus the amount of losses is low. Another point is that losses exceeding 20 % never permit retention time to increase above 12 years, independently of the amount of TBV removed. This relationship is clear in Fig. 6.5, where the RT is a variable dependent on evaporation at different levels of TBV removal.

Figure 6.6 shows the overall relationship between the variables considered (TBV production, yield, and evaporation rate). It is worth noting that the highest RT is only possible when the yield is high (low losses/evaporation) and TBV removal is low. Decreasing yield and increasing TBV removal reduce the RT.

The application of the theoretical model described above provides valuable information on the management of a barrel set. Furthermore, with two simple and easily measurable parameters (TBV tapped off and BVR refilling), it is possible to

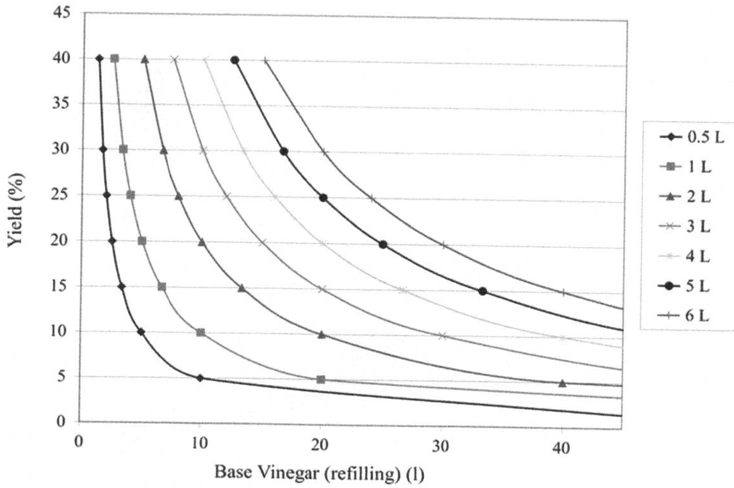

Fig. 6.3 Relationship between yield and refilling volume of a typical barrel set of TBV

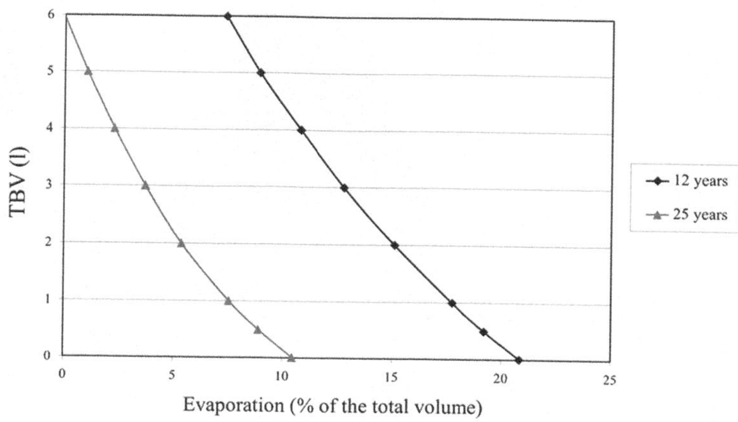

Fig. 6.4 Volume of TBV tapped off as a function of evaporation (expressed as a % of the volume of the barrel set) at two different levels of RT

correctly estimate the RT of the TBV removed. Both the yield and the RT are significantly affected by the evaporated fraction and more in general by all product losses.

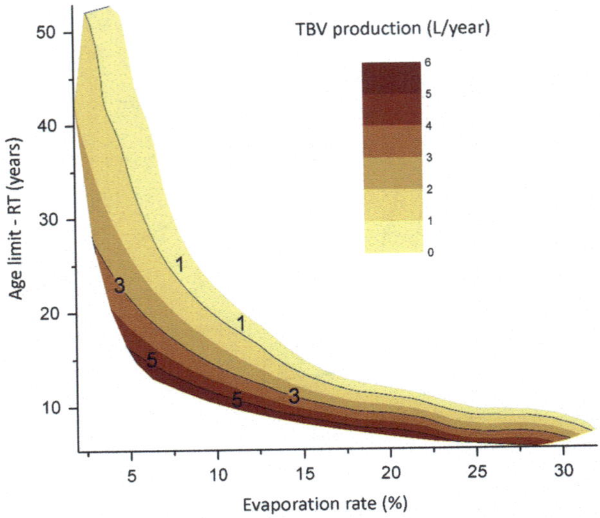

Fig. 6.5 Residence time as a function of evaporation at different levels of TBV production

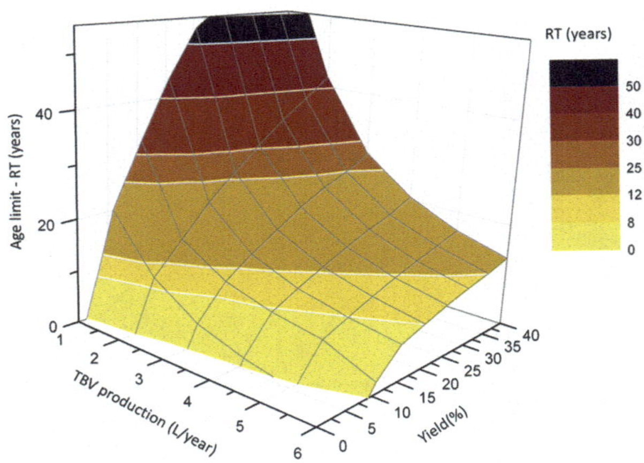

Fig. 6.6 The effect of yield and TBV removal on the residence time of TBV

6.1.1 *The Mass Model*

The volume model described above is based on the assumption that the vinegar inside the barrels has the same composition during storage. This is a necessary assumption for a simplified model in order to solve the problem of blending due to the refilling procedure in the barrel set.

A more precise assessment of the RT of vinegar should take into account the mass of solute rather than the volume of the vinegar, since a consistent loss of water occurs during storage. As previously reported, the yield is the ratio between the amount of vinegar removed annually and the amount of base vinegar put into the barrel set at the equilibrium of the system.

The outputs, in volume, are represented by:

- Evaporation (V_e) through the wood and the holes on top of the barrels
- The vinegar removed (V_{tbv})
- Any product losses occurring from the barrels (V_p)

The sum of the output volumes is exactly compensated by the refilling volume (V_r) with new base vinegar, so:

$$V_r = V_e + V_{tbv} + V_p. \tag{6.15}$$

Yield by volume is therefore:

$$\text{Yield}(v) = \frac{V_{tbv}}{V_r} \times 100. \tag{6.16}$$

The yield by weight is obtained by multiplication of the volumes of the TBV and refilling by their relative density values, respectively, d_{tbv} and d_r:

$$\text{Yield}(w) = \frac{V_{tbv} \times d_{tbv}}{V_r \times d_r} \times 100. \tag{6.17}$$

The substitution of the volumes with the masses in the previous model gives:

$$E_i(n) = \frac{M_i^{\text{residual}}(n)}{M_i^{\text{final}}(n)} \cdot [E_{i+1}(n-1) + 1] + \frac{M_i^{\text{refilling}}(n)}{M_i^{\text{final}}(n)} \cdot [E_i(n-1) + 1]. \tag{6.18}$$

On the basis of barrel capacity, refilling volumes, amount of TBV removed, concentration of solutes in the base vinegar, and the ageing yield, the factors of the following equations can be calculated in recursive manner.

- $M_i^{\text{final}}(n) = M_i^{\text{residual}}(n) + M_i^{\text{refilling}}(n)$ (mass of the solutes after refilling)
- $M_i^{\text{residual}}(n) = M_i(n-1) - M_i^{\text{lost}}(n)$ (mass of the solutes before refilling)
- $M_i^{\text{removed}}(n) = \left[M_i^{\text{residual}}(n)/V_i^{\text{residual}}(n)\right] \times V_{i-1}^{\text{refilling}}(n)$
- $V_i^{\text{residual}} = V_i - V_i^{\text{evaporated}}$
- $V_i^{\text{refilling}}(n) = V_i^{\text{evaporated}} + V_{i-1}^{\text{refilling}}(n)$
- $M_i^{\text{refilling}}(n) = M_{i+1}^{\text{removed}}(n)$

Figure 6.7 shows the residence time trend for the vinegar in each barrel and for the TBV removed. The continuous lines are the results from the new model based on mass, and the dashed lines refer to the previous model based on volume. During the initial period of maturing, the differences are negligible. When the age limit is reached, in the case of the study, the residence time by volume is 18 % lower than by mass, this discrepancy being greater in barrels containing the oldest vinegar.

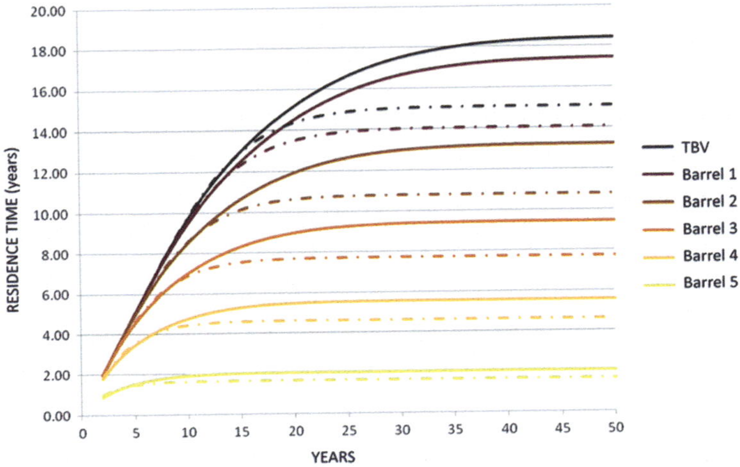

Fig. 6.7 Simulation of the residence time of vinegar in a typical barrel set of TBV. The *continuous lines* are the results from the model based on mass, and the *dashed lines* refer to the model based on volume

6.2 The Yield, the Real Result

The yield of the ageing process derives from a comparison of the total mass of solutes entering the barrel set with the mass of the removed TBV. The barrel set is in steady state when it produces TBV at constant density throughout the years. In such conditions of equilibrium, the yield of the ageing process will be constant but always less than 100 % due to the losses of the system. The mass of solutes introduced with the base vinegar is equivalent to the mass of solutes of the TBV removed, plus the losses that occur during the transition through the barrel set.

Many phenomena can contribute to the loss of solutes during ageing. The physical losses include the evaporation of volatile substances (alcohols, esters, and organic acids), leakage due to structural defects in the casks, and the formation of precipitates. From a chemical point of view, TBV is rich in carbonyl compounds and amino acids able to react in condensation reactions, such as aldol condensation and the Maillard reaction. These reactions lead to the expulsion of small molecules like water, CO_2, and small aldehydes and so contribute to the loss of solutes. Finally, biological processes can lead to further consumption of solutes. For example, when the concentration of sugars is not sufficient to inhibit the activity of bacteria, biological oxidation of acetic acid can occur.

The assessment of the overall losses from a barrel set during ageing is difficult because a large number of processes of different nature are involved. However, reliable values were empirically obtained from various barrel sets that have been studied over the years.

The best way to establish the exact mass of solutes of TBV is to measure the water content. However, an approximated value can be calculated on the basis of the relative density, the concentration of reducing sugars (determined with Fehling assay), and with the aid of the National Bureau of Standards (NBS, now NIST) saccharimetry tables (Circular of the NBS C440, Table 115).

Abbreviations:

- d = relative density
- $[z]$ = concentration of reducing sugars (g/100 ml)
- $m(z)$ = mass of reducing sugars in 100 ml of solution (g)
- $m(w)$ = mass of water in 100 ml of solution (g)
- $v(w)$ = volume of water in 100 ml of solution (ml) = $m(w) \times d(w)$
- $m(x)$ = mass of solutes other than reducing sugars (g)
- $v(z)$ = volume of reducing sugars in 100 ml (ml)
- SV(z) = specific volume of sugars in solution at concentration $[z]$ (ml/g)

Approximations:

- Specific volume of sucrose is equal to specific volume of reducing sugars
- Relative density of water is equal to 1
- The volume of solutes other than reducing sugars is negligible

The total mass of solutes in 100 ml of vinegars is:

$$d(v) \times 100 = m(z) + m(w) + m(x). \tag{6.19}$$

The sum of the volume of water and sugars is:

$$100 = v(z) + m(w), \tag{6.20}$$

where:

$$v(z) = \mathrm{SV}(z) \times m(z). \tag{6.21}$$

From Eq. (6.19) an approximated value of $v(w)$ is obtained.

$$v(w) = m(w) = (d(v) \times 100) - m(z). \tag{6.22}$$

From the proportion: $m(z){:}v(w) = m100(z){:}100$, extract the value of $m(z)$ in 100 ml of water: $m100(z) = (m(z) \times 100)/v(w)$.

Using the saccharimetry tables, extract, by linear interpolation, the volume of $m100(z)$ g of sucrose at concentration $[z]$ and then the specific volume of sugars at the same concentration: SV(z).

From Eq. (6.21), substitute $v(z)$ in Eq. (6.20) and obtain $m(w)$, then from Eq. (6.19) obtain $m(x)$.

The sum of $m(z)$ and $m(x)$ is a good approximation of the total solids in 100 ml of vinegar, so this approximation was used to establish the input and output masses of

Table 6.1 Yield of two barrel sets on the basis of mass of solutes

Barrel set	Mass of solutes in BVR (g/100 ml)	Volume of BVR (l)	Mass of input solutes (g)	Mass of solutes in TBV (g/100 ml)	Volume of TBV (l)	Mass of output solutes (g)	Yield (%)	Loss of solutes (%)
A	20.4	22.25	4,539	65.1	4.00	2,604	57	43
B	20.4	16.13	3,290	67.6	4.00	2,704	82.2	17.8

Table 6.2 Overall scheme of solutes loss, from grapes to balsamic vinegar, split into technological steps

Description	Substrate (solutes concentration)	Yield (% by weight)	Product (solutes concentration)	Solutes concentration change (%)
Pressing of grapes				
	Grapes		Fresh grape must	
		100[a]	21.6 g/100 ml; 20°Bx	0
Cooking of must				
Water evaporation Peptides coagulation Tartrates precipitation	Fresh grape must (21.6 g/100 ml; 20°Bx)	97–99[b]	Cooked must (33.8–47.1 g/100 ml; 30–40°Bx)	+ (56–118) + (50–100)
Alcoholic fermentation				
Loss of CO_2 Removal of cells Tartrates precipitation	Cooked must (33.8–47.1 g/100 ml; 30–40°Bx)	35–40 65–80[c]	Base wine (8 % v/v, 6.24 g/100 ml EtOH; 22.2–37.7 g/100 ml)	− (20–34)
Acetic acid fermentation				
Suroxidation Oxidation of organic substances	Base wine (8 % v/v, 6.24 g/100 ml EtOH; 22.2–37.7 g/100 ml)	Max 97	Base vinegar (6.05 g/100 ml AcOH; 22.0–37.5 g/100 ml)	− (0.5–0.9)
Ageing				
Evaporation of water and volatile substances Formation of precipitates Chemical reactions Structural leakages	Base vinegar (22.0 g/100 ml)	55–85	Traditional balsamic vinegar (65.1–67.6 g/100 ml)	+ (196–207)

[a]100 % as regards the solutes, while the crushing and pressing yield is max 70 %
[b]Assessed on the basis of the amount of residual tartaric acid and the concentration of fructose after cooking (Falcone and Giudici 2010)
[c]Effective yield obtained considering that only part of the sugars is converted to ethanol

a barrel set. The average values of input and output masses per year for two barrel sets are reported in Table 6.1. The yield and the loss of solutes are quite different between the two barrel sets; the losses of solutes range between 15 and 45 % (corresponding to a yield of 55–85 %).

On the basis of the experimental data reported in literature (Lemmetti and Giudici 2010a, b; Lemmetti 2013), a table was created summarising all the yields of the different production steps, with the purpose of calculating how many kg of grapes are required to produce 1 kg of TBV (Table 6.2).

The overall yield, expressed as the ratio between the solutes of TBV recovered and the amount of solutes introduced into the system as grape must, ranges from 34 to 65 %; this means that, to obtain one litre of TBV at a concentration of about 66 % (w/v) of soluble solids, a maximum of 10 kg of must at 20 % (w/v) of solutes is required for the least favourable yield, and only 5 kg of grape must is necessary when the yield is high.

In order to be consistent with the data collected from real barrel sets, the yield range clearly shows that the amount of solutes lost in the manufacturing process can be significantly reduced with economic and qualitative advantages. In fact, the yield, as proven above, significantly influences the residence time of TBV.

6.3 Maturity: Physical and Chemical Modification of the Vinegar Constituents

In the introduction to the chapter, a clear distinction was made between RT and maturity during the ageing processes of TBV. The first is the time required for turnover, and the second is the time-dependent changes in chemical, physical, and sensorial properties occurring during RT.

Several studies have attempted to correlate the composition of TBV and its ageing time (Anklam et al. 1998; Baroni et al. 2009; Blasi et al. 2002; Chiavaro et al. 1998; Cirlini et al. 2009, 2011; Cocchi et al. 2004; Consonni and Cagliani 2007; Consonni et al. 2008; Del Signore 1996, 2001; Del Signore et al. 1998; Giaccio et al. 1996; Giudici 1993; Giudici et al. 1994; Zeppa et al. 2002). The most common target molecules in these studies are organic compounds present in solution and/or the volatile fraction, such as alcohols, ethers, carbonyl compounds, amino acids, sugars, and heterocyclic compounds, along with trace metals. Some studies focused on specific molecules, in particular, furan derivatives (Masino et al. 2005), because these compounds are known to be the first products of the browning reaction due to degradation of carbohydrates.

The conclusions of all these chemical determinations are generally still unsatisfactory due to the complexity of the production process, the lack of information about the properties of the raw materials used over the years, and more importantly, the certainty of the RT of the samples examined.

The first experimental evidence of a slowing down of chemical conversion during TBV ageing toward an RT of 14 years was provided by the present authors (Falcone and Giudici 2008, 2009). They demonstrated that polymerization reactions are first activated during the cooking of the grape must and then progress more slowly over the years. Recently, more studies have been conducted to characterise melanoidins from a chemical and spectroscopic point of view (Lemmetti 2013).

Viscosity measurements (dynamic and kinematic) can also provide relevant information on the different types of balsamic vinegar. The rheological properties

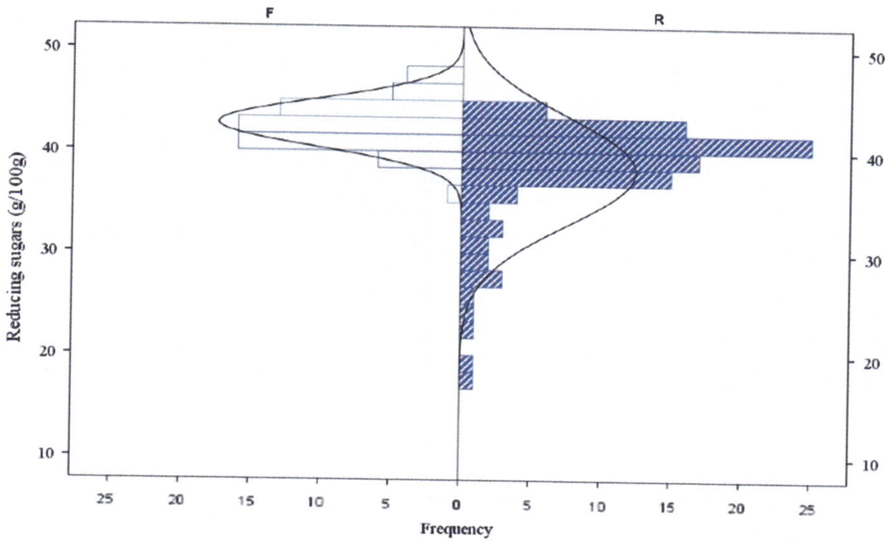

Fig. 6.8 Concentration of sugars in TBV samples with different RT. On the *right* 100 on-going TBV samples, RT less than 12 years. On the *left* 60 matured TBV samples, RT more than 12 years

of traditional balsamic vinegar have been thoroughly studied with the aim of defining a model for quality assessment (Falcone et al. 2007, 2008).

6.3.1 *Chemical Changes*

A large number of TBVs, characterised by different RT, including 60 samples with more and 100 with less than 12 years of RT, were analysed, considering several macroscopic components: sugar, titratable acidity, acetic acid, Brix degree, and pH. The results are reported below (Figs. 6.8, 6.9, 6.10, and 6.11).

Generally, cooked must for TBV has a sugar concentration between 30 and 40 %. Subsequently, this value decreases due to alcoholic fermentation and direct oxidation of glucose by acetic acid bacteria. On the other hand, ageing tends to progressively increase the concentration of sugars up to 50 %. It should be noted that the amount of reducing sugars is not able to justify the high values of specific weight encountered in TBV samples. The discrepancy between the expected and the observed values of relative density indicates the presence of solutes (derived from sugar) undetectable by titration with Fehling's reagent. Moreover, this is an indirect confirmation of the profound changes the sugars undergo during the slow ageing process.

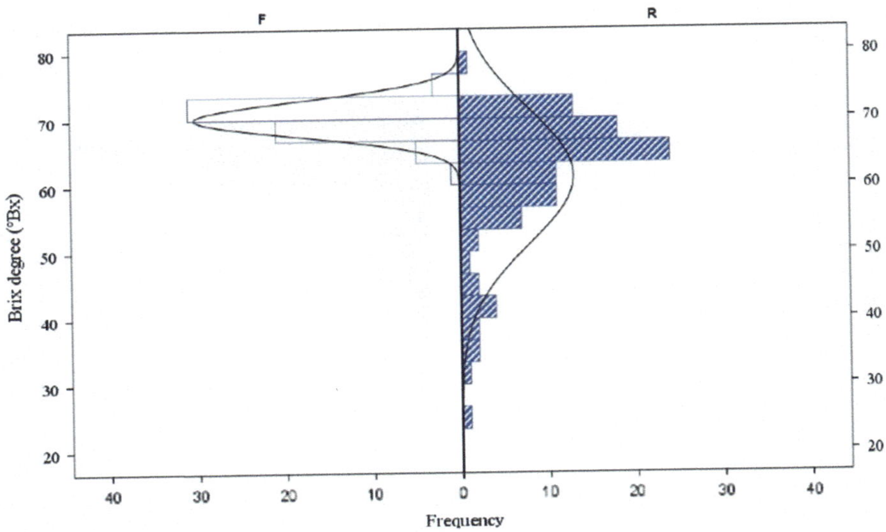

Fig. 6.9 Soluble solids in samples of TBV, year 2010. On the *right* 100 on-going TBV samples, RT less than 12 years. On the *left* 60 matured TBV samples, RT more than 12 years

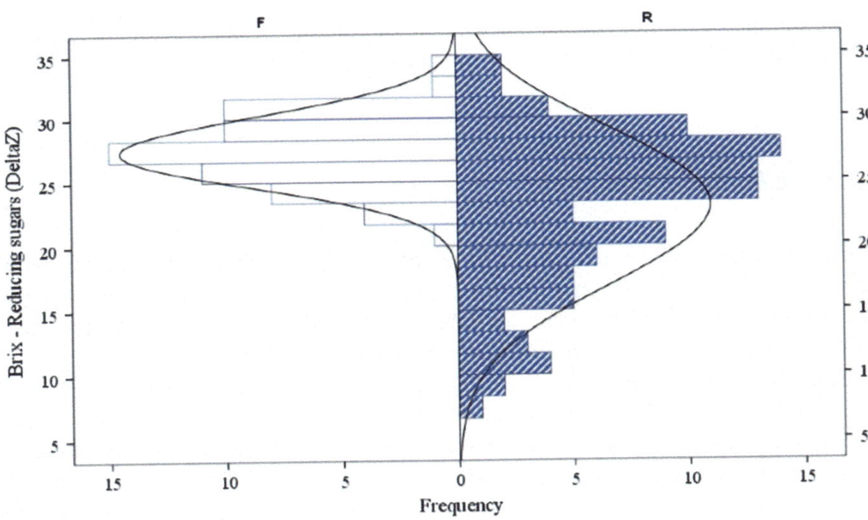

Fig. 6.10 Distribution of DeltaZ (algebraic difference between Brix degree and reducing sugars content) in the samples of TBV, year 2010. On the *right* 100 on-going TBV samples, RT less than 12 years. On the *left* 60 matured TBV samples, RT more than 12 years

Similar observations can be made for the Brix value, which is completely different from the expected value based on the concentration of sugar. This difference is especially large for extensively aged products.

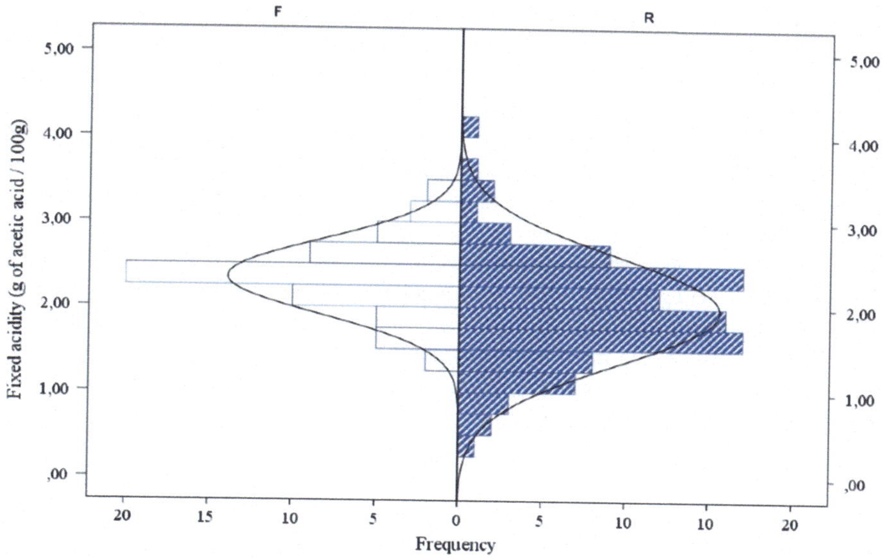

Fig. 6.11 Distribution of fixed acidity in the samples of TBV, year 2010. On the *right* 100 - on-going TBV samples, RT less than 12 years. On the *left* 60 matured TBV samples, RT more than 12 years

Furthermore, in BV, BC, and BVM, which mainly contain simple sugars, the Brix value is only slightly higher than the sugar content. In samples of TBV with more than 12 years of RT, the Brix values are frequently higher than 70 (Fig. 6.9) due to the large number of compounds that can cause interference, as well as the marked degradation of simple sugars, which are converted into non-reducing species. The high-molecular weight compounds from the last stages of the Maillard reaction are the perfect candidates to explain the lack of sugars observed. In fact, the thermal treatment of must triggers the process of non-enzymatic browning, with the formation of compounds which can evolve towards a large number of chemical species, such as furfurals and the so-called advanced glycation end products (AGEs), while other species are able to participate in the formation of high-molecular weight compounds (melanoidins) that arise during ageing.

The difference, in dimensionless value, between Brix degree and reducing sugars is defined as DeltaZ and was postulated to be related to the level of degradation, modification, and polymerization of sugars (Lemmetti and Giudici 2011; Giudici and Falcone 2010). Figure 6.10 shows the DeltaZ of aged vinegars (more than 12 years RT) and younger samples. It is clear that, from the distribution in the sample population, the TBV's with longer RT tend to have higher DeltaZ. In high-quality TBV samples, the DeltaZ is greater than 20 and it can rise above 30 units (see Chap. 3).

During the long ageing period, vinegar concentrates due to the evaporation of water and other volatile substances up to 70 % by weight of solutes. It is noteworthy that the ratio between the total amount of solutes and sugars during this stage is not predictable on the basis of the concentration process only. Total concentration of soluble solids increases by a factor of 2 or 3, but the amount of acids and other organic compounds (other than sugars) does not usually increase more than 10 %. Considering the concentration, it would be reasonable to expect a great sugar content in highly concentrated and aged TBV. The most appreciated TBV's do exhibit high contents of soluble solids (>70 %) but only exhibit a maximum of 45 % of residual sugars (Lemmetti and Giudici 2012).

The analysis of TBV samples indicates that the minimum value for fixed acidity is equal to 1.6 % (in grams of acetic acid per 100 g), corresponding to at least 2.4 % by volume. Compared to other balsamic products, the fixed acidity of TBV is quite high. For this reason, the fixed acidity is another strongly indicative parameter for discriminating TBV from other products made by mixing concentrated must and wine vinegar, without a maturing process.

6.3.1.1 Sugar Degradation and High-Molecular Weight Compounds (Melanoidins) in TBV

All foods containing simple or complex sugars and amine or amide group, exposed to cooking or roasting processes, undergo non-enzymatic browning reactions (Maillard reactions) with formation of a wide range of compounds, including the so-called polymeric melanoidins. These compounds affect the aroma, colour, flavour, and rheology of foods, as well as having antioxidant, antimicrobial, anti-hypertensive, and probiotic properties. Recently, the presence of high-molecular weight compounds in TBV has been established (Lemmetti 2013). Paper filtration and ultrafiltration membranes separated fractions of different sizes (in kDa), as reported in Table 6.3.

The quantity of solutes recovered from each fractionation step is not proportional to the concentration of the samples and RT. Several factors were assumed to be significant, including the handling of the barrel, the amount of precipitate, and the stratification of the product. It was also postulated that the solubility of these polymeric substances reaches a limit, excluding indefinite stockpiling of melanoidins in solution.

The pH of TBV is relatively low (less than 3), which could have an important influence on this phenomenon, since the isoelectric point of some classes of melanoidins is close to 2.5. The solubility of melanoidins is also influenced by the presence of metal cations that act as flocculation agents, a condition that can occur in TBV.

The high-molecular weight fractions (HMW, Ax_10k, and Ax_3k) were identified as melanoidins on the basis of their absorption spectra in UV and visible light. Melanoidins strongly absorb in the 260–290 nm region, often without a clear maximum but in the form of a plateau (Fig. 6.12).

The infrared (IR) spectra of all these high-molecular weight fractions are nearly identical and independent of their size. Therefore, it was assumed that the building

Table 6.3 Weight of fractions obtained from simple paper filtration and ultrafiltration of five TBV samples of decreasing maturation time

Sample	Soluble solids (°Bx)	Reducing sugars (g/100 g)	Weight (mg)	Insoluble matter (mg–% by weight)	Fraction >10 kDa (mg)	Fraction 3–10 kDa (mg)	Fraction < 3 kDa (mg)
A1	63.8	46.2	1,006.5	70.1–7.0	138.8	89.7	932.6
A2	57.4	39.0	994.3	115.3–11.6	197.7	147.3	1,024.3
A3	52.4	38.1	1,009.1	77.2–7.6	157.5	94.4	924.4
A4	44.8	30.9	1,001.2	83.1–8.3	172.5	176.8	1,038.2
A5	37.7	25.9	1,006.0	55.7–5.5	102.3	101.0	789.7

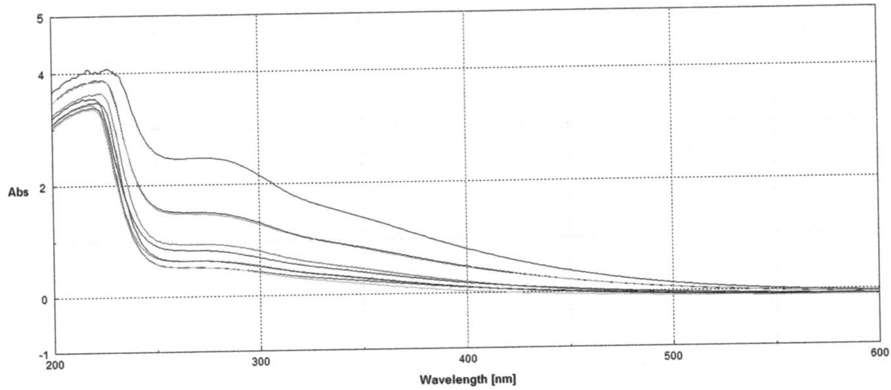

Fig. 6.12 Absorption spectra (200–600 nm) of high-molecular weight fractions of TBV samples

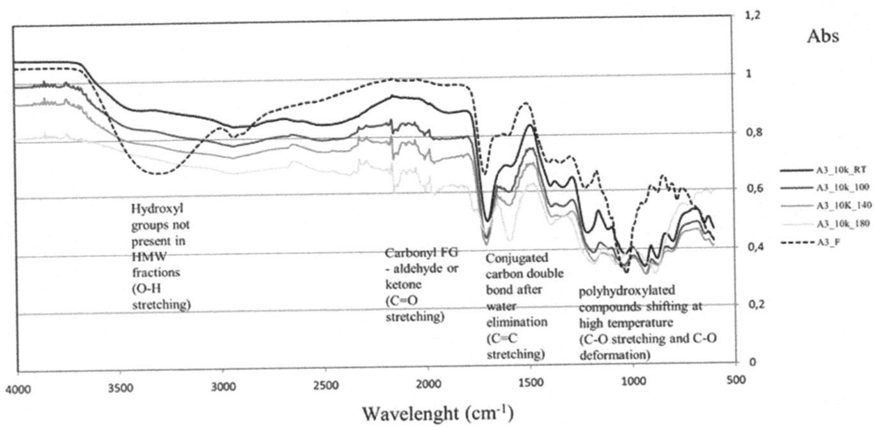

Fig. 6.13 IR absorption spectra of the fractions from sample A3. A3_F is the filtrate fraction, with the lowest molecular weight compounds, A3_10k is the highest weight. Heating of A3_10k fraction causes dehydroxylation

blocks of melanoidins are formed during the first stages of the technological process, the cooking of the must, and polymerization occurs during the maturation stage, when the increasing loss of water favours condensation reactions.

The IR spectra of high-molecular weight fractions share many absorption bands with the spectra of the residual fraction (Ax_F), which is low molecular weight (<3 kDa), mainly constituted by monosaccharides. Figure 6.13 shows the absorption spectra of the sample A3, in particular, the low-molecular weight fraction (A3_F) and the high-molecular weight fraction (A3_10k) exposed to heating up to 180 °C.

From this comparison, some information on the chemical structure of melanoidins was delineated. Fraction A3_F shows some of the typical absorption bands of carbohydrates and carboxylic acids: a wide band (O–H stretching, 3,100–3,600 cm^{-1}) due to the presence of residual water and associated hydroxyl functional groups, aliphatic signals (C–H stretching, 2,910–2,960 cm^{-1}), carbonyl compounds (C=O stretching, 1,720 cm^{-1}), and combination of C–O stretching and deformation (1,040–1,055 cm^{-1}) referable to polyhydroxylated compounds (Rubinsztain et al. 1986).

The spectra are not well resolved. Therefore, even if there are other characteristic signals, the exact nature of the carbonyl signal was not ascertained. The set of signals from simple sugar fractions A3_F has the same spectra at room temperature as the high-molecular weight fraction (A3_10k_RT) which, as verified by size-exclusion chromatography, does not contain monosaccharides or other small organic molecules. The only difference is the absence of the absorption band of hydroxyl groups. It was postulated that these functional groups are, in some way, involved in interactions that only occur in high-molecular weight compounds.

A further attempt to identify the melanoidins was performed by NMR on purified gel filtration fractions. It is significant that all the spectra of high-molecular weight compounds contain carbohydrate signals and no aromatic moieties. The HSQC spectra in Fig. 6.14 show some broad bands [3.2–4.3 ppm (^{1}H); 60–85 ppm (^{13}C)] from carbohydrates non-anomeric signals and [4.4–5.1 ppm (^{1}H); 102–103 ppm (^{13}C)] due to anomeric centres. In addition, alkyl chain signals are also present.

Moreover, the high-field multiplet signal [1.3 ppm (^{1}H); 18 ppm (^{13}C)] comes from alkyl protons that correlate with some protons of the carbohydrate moieties (Fig. 6.15). This was postulated as due to the formation of glycosidic bonds from the hemiacetal form of saccharides with alcohols or other nucleophilic species.

In conclusion, based on the data of Lemmetti (2013), there is strong evidence that the TBV melanoidins comprise a skeleton of carbohydrate moieties with small amounts of nitrogen, as previously reported by the present authors (Falcone and Giudici 2008).

6.3.2 *Physical Change*

Development of an instrumental test for fluid viscosity measurements requires knowledge about its rheological behaviour under flow as well as its characteristic relaxation time. Relaxation time is a characteristic property of fluids and provides an estimate of the time required for relaxation of the stresses arising from applied flow rates.

Newtonian fluids are ideal fluids exhibiting a dissipative behaviour which means they relax stresses instantaneously when the flow stops, and they do not have memory of the original molecular configuration. The characteristic relaxation time for a Newtonian fluid is zero, and its viscosity, i.e. its resistance to flow, increases proportionally with the flow rate but does not depend on the timescale of

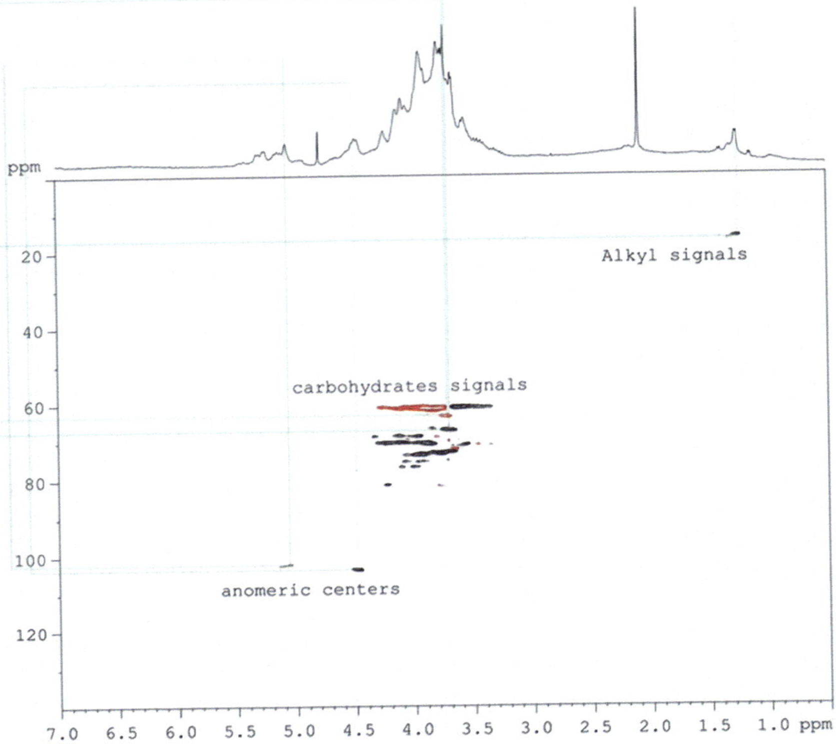

Fig. 6.14 HSQC spectra of fraction A1_10k. In addition to carbohydrate signals, alkyl chain signals are also present

the flow experiment. Newtonian viscosity can easily be calculated as the ratio between the applied flow rate and its corresponding shear stress.

The flow of many real fluids differs from the Newtonian model, exhibiting a conservative behaviour (viscoelastic flow). These fluids have a finite relaxation time and partially recover the initial molecular configuration when the flow stops. Scientific literature reports a plethora of viscoelastic flow behaviours (Batchelor 2000). For example, a pseudoplastic or shear-thinning flow can be observed when fluid viscosity decreases with the flow rate, a shear-thickening or shear-dilatant flow can be observed when the viscosity increases with flow rate, and thixotropic or rheopectic flow can be detected where the viscosity increases or decreases respectively with time.

The knowledge of the ratio between the characteristic relaxation time and the timescale of the experiment represents the first step in flow analysis, making it possible to establish whether or not the effects of relaxation dominate in a given instrumental test. For example, a viscoelastic fluid behaves like an elastic solid when a flow is imposed over very short timescales; in this case, viscosity increases

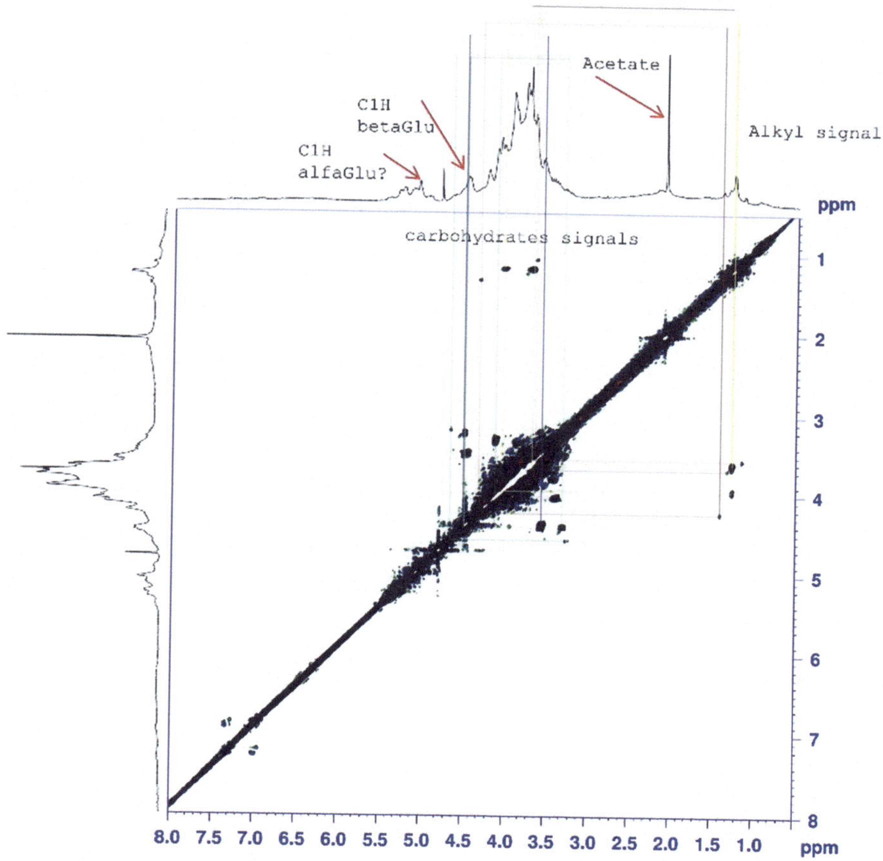

Fig. 6.15 H-H COSY spectrum of fraction A1_10k. Alkyl signals correlate with protons in the carbohydrates region

towards infinity. Conversely, it behaves like viscous liquid when a flow is imposed over very long timescales. For an intermediate timescale, the resistance to flow depends on the flow rate or the timescale of the flow experiment. For a viscoelastic fluid, only an apparent viscosity can be estimated as the ratio between flow rate and corresponding shear stresses. An instrumental test capable of accurately determining fluid viscosity requires keeping the sample at rest for a sufficient period of time to relax all stresses induced during sample handling before the flow analysis.

Recently, an accurate instrumental method capable of characterising the rheological behaviour and determining the viscosity of TBV was published (Falcone et al. 2008). In this experimental work, vinegar samples were collected during a public competition organised by the Confraternity of TBV of Reggio Emilia, aimed at classifying more than 1,000 vinegars on the basis of their sensorial quality. Using a shear-controlled rheometer (TA Instruments, Denmark), the authors analysed the flow behaviour of the top 20 ranked vinegars (indicated below as "F") as well as another 100 chosen randomly from among all the vinegars participating in the sensorial competition (indicated below as "R").

All the vinegars investigated were characterised according to their RT. They were also characterised for some basic properties in order to evaluate the different degrees of chemical vinegar maturity.

Results from flow analysis indicated a wide range of rheological responses, ranging from Newtonian to pseudoplastic behaviour. Authors selected 60 and 600 s^{-1} as bottom and upper limits for the flow rate at which the apparent viscosity should be measured and proposed their mathematical difference (indicated as DeltaV) to link the observed rheological behaviour to the degree of vinegar maturity.

For the sake of clarity, Fig. 6.16 shows the rheological behaviour of both R and F vinegar samples as a function of their viscosity and DeltaV property. Positive values for the adapted DeltaV property indicate the deviation of flow behaviour from Newtonian towards shear thinning (viscosity increases with shear rate). Negative DeltaV values indicate shear-thickening behaviour (viscosity decreases with shear rate). Interestingly, these data infer that the highest DeltaV values are estimated for the top-ranked vinegars, while only two of these 20 vinegars have a negative value. Furthermore, the vinegars with lowest viscosity exhibit the widest deviation from Newtonian behaviour.

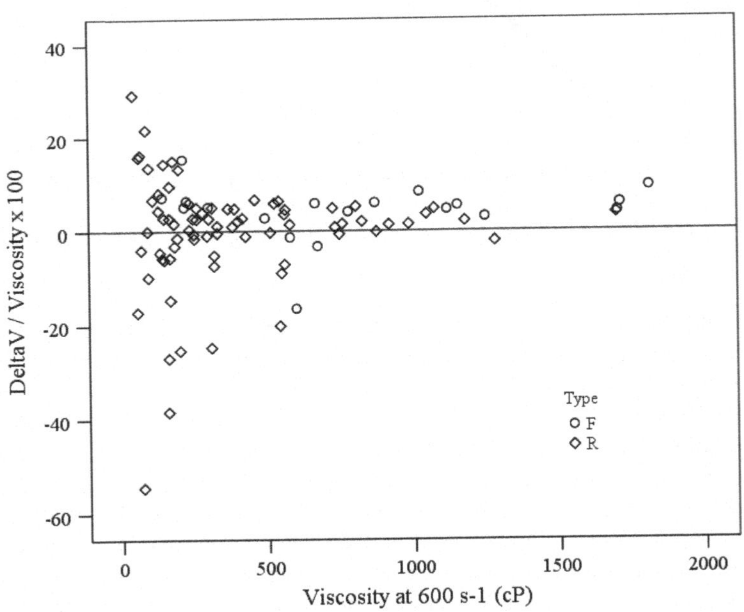

Fig. 6.16 DeltaV values around zero indicate Newtonian behaviour. Positive DeltaV indicates pseudoplastic flow behaviour. *Circle markers* (F) are the best quality TBV samples, and *diamond markers* (R) are randomly chosen vinegars

The high variability of observed flow behaviour between the investigated vinegars can be attributed to the complexity of both the vinegar composition and the physical state of their components, arising from different technological procedures as well as different degrees of vinegar maturation.

Recent studies provide evidence that prolonged ageing of TBV inside wooden barrels allows crystallisation of the alpha-glucose monohydrate, as well as allowing the sugar degradation reactions to progress to such an extent that amorphous and highly dispersed polymeric structures (known as melanoidins) accumulate, self-assemble, and form colloids suspended in the vinegar bulk or form gel-like network structures, the latter eventually expanding throughout the entire volume.

The presence of these polymers strongly influences the rheological behaviour of the vinegars; gel-like networks exhibit viscoelastic flow, while dispersed colloids may induce shear-thinning (pseudoplastic) or shear-thickening flow (Falcone 2010; Falcone and Giudici 2008; Falcone et al. 2008, 2011, 2012; Lemmetti et al. 2013).

Dynamic flow experiments were recently carried out in order to highlight the viscoelastic behaviour of partially solidified vinegars. A pseudoplastic behaviour is most frequently observed during flow analysis of TBV, while shear-thickening flow is rarely observed. For example, Fig. 6.17 reports the flow curve of a random sample showing rheological shear thickening.

Shear-thickening flow is characterised by an increase in apparent viscosity with flow rate. Such behaviour is typical of polymeric solutions in which the alignment

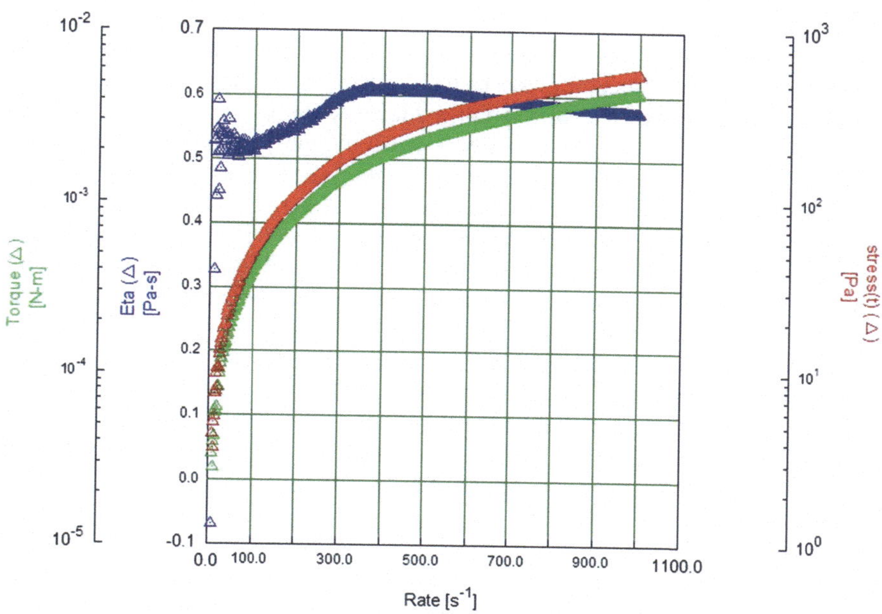

Fig. 6.17 Flow curve of a random TBV sample showing rheological shear-thickening flow behaviour

of the molecular chains increases with the flow rate, resulting in an increase in hydrogen or polar bonds.

Unfortunately, no studies have been carried out to date to prove whether these polymeric hydrocolloids are naturally formed or intentionally added during vinegar production in order to increase viscosity. It is reasonable to hypothesise that in order to reduce the time required for vinegar maturation, a wide range of food-grade hydrocolloids could be used as very effective shear-thickening agents in vinegar production. Any such practice must be considered as commercial fraud.

The main factor causing pseudoplastic behaviour in TBV has been attributed to the breaking effect of hydrodynamic forces on the colloidal melanoidins accumulated during vinegar maturation. The degree of shear thinning depends on both the volume fraction of the dispersed melanoidins and the composition of the vinegar bulk fraction. Evidence was provided that the accumulation of polymeric and colloidal melanoidins progresses with the time of residence of vinegar in the barrel set, not approaching an upper limit before at least 14 years of RT (Falcone and Giudici 2008, 2009).

Besides its usefulness for characterising the quality of vinegars, flow data can also be used to reveal the applied vinegar technology, ageing conditions, and susceptibility to solidification phenomena (physical stability). Falcone et al. (2007) investigated the degree of maturity and rheological behaviour of a set of TBV of Modena collected during a sensorial competition.

As can be inferred from Fig. 6.18, the vinegars can be classified according to their degree of maturity using only four basic properties, including DeltaV (in this case measured at shear rates of 500 s^{-1} and 10 s^{-1}), soluble solids content (expressed in Brix degree units), concentration of reducing sugars (assessed according to Fehling's method), and DeltaZ (calculated as the difference between the Brix degree and the concentration of reducing sugars).

DeltaV assumes a value close to 1 only when the vinegars behave as a Newtonian fluid. In this case, the application of higher shear rates does not produce structure variations or decrease or increase the viscosity. It is worth noting that the vinegars exhibiting Newtonian flow behaviour were the only vinegars that undergo sugar degradation processes and water evaporation processes without problems of solidification from ageing. For these vinegars, both the Brix and DeltaZ increase, but DeltaV remains close to 1.

Samples in which DeltaV produces a value less than 1 represent vinegars characterised by the presence of suspended colloids or gel-like aggregates. In these vinegar samples, Brix increases with DeltaV, but the increase in DeltaZ values does not correspond to an increase of DeltaV. These results indicate that the water evaporation processes occurred over an excessively short time during vinegar production.

In conclusion, a knowledge of the apparent viscosity measured at low- and high-flow rates (DeltaV), the Brix degree, and the concentration of reducing sugars makes it possible to classify balsamic vinegars according to the degree of maturity.

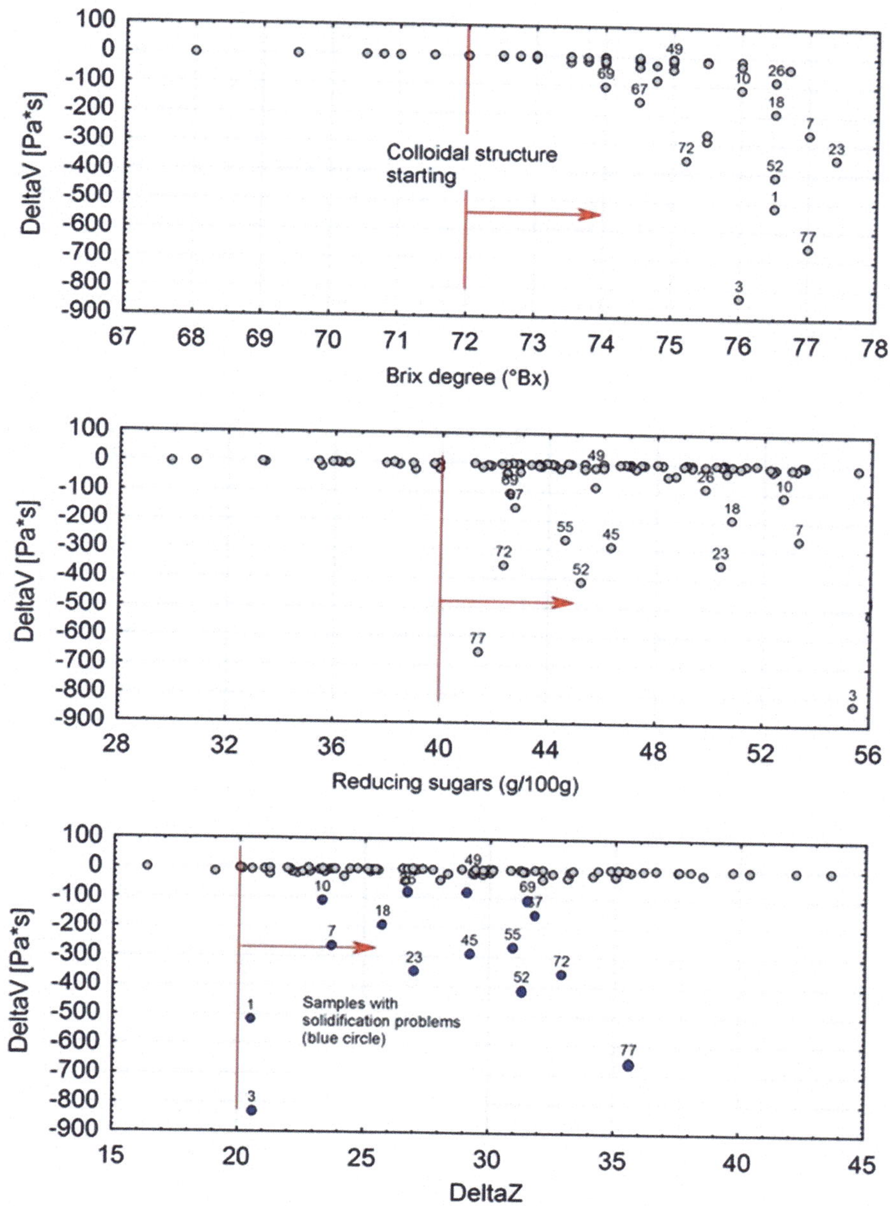

Fig. 6.18 Measurements of viscosity, Brix, and sugar concentration for use as indicators of physical stability and authenticity of the process

References

Anklam E, Lipp M, Radovic B, Chiavaro E, Palla G (1998) Characterisation of Italian vinegar by pyrolysis–mass spectrometry and a sensor device ('electronic nose'). Food Chem 61 (1–2):243–248

Baroni S, Consonni R, Ferrante G, Aime S (2009) Relaxometric studies for food characterization: the case of balsamic and traditional balsamic vinegars. J Agric Food Chem 57(8):3028–3032

Batchelor GK (2000) An introduction to fluid dynamics, Cambridge Mathematical Library series. Cambridge University Press, Cambridge

Blasi F, Riccieri R, Brutti M et al (2002) HRGC-MS of free enantiomeric amino acids in aged balsamic vinegars from Modena. J Commodity Sci 41:333–345

Chiavaro E, Caligiani A, Palla G (1998) Chiral indicators of ageing in balsamic vinegars of Modena. Ital J Food Sci 10:329–337

Cirlini M, Caligiani A, Palla G (2009) Formation of glucose and fructose acetates during maturation and ageing of balsamic vinegars. Food Chem 112:51–56

Cirlini M, Caligiani A, Palla L, Palla G (2011) HS-SPME/GC-MS and chemometrics for the classification of Balsamic Vinegars of Modena of different maturation and ageing. Food Chem 124:1678–1683

Cocchi M, Durante C, Foca G et al (2004) Application of a wavelet-based algorithm on HS-SPME/GC signals for the classification of balsamic vinegar. Chemometr Intell Lab 71:129–140

Consonni R, Cagliani LR (2007) NMR relaxation data for quality characterization of Balsamic vinegar of Modena. Talanta 73:332–339

Consonni R, Cagliani LR, Benevelli F et al (2008) NMR and chemometric methods: A powerful combination for characterization of balsamic and traditional balsamic vinegar of Modena. Anal Chim Acta 611:31–40

Del Signore A (1996) Differentiation of balsamic vinegars using physical indexes. J Commodity Sci 37:51–56

Del Signore A (2001) Chemometric analysis and chemical and physical parameters of traditional balsamic vinegars from Modena. J Commodity Sci 40:33–48

Del Signore A, Campisi B, Di Giacomo F (1998) Characterization of balsamic vinegar by multivariate statistical analysis of trace element content. J AOAC Int 81:1087–1095

Falcone PM (2010) Crystallization and jamming in the traditional balsamic vinegar. Food Res Int 43:2217–2220

Falcone PM, Giudici P (2008) Molecular size and molecular weight distribution impacting traditional balsamic vinegar ageing. J Agric Food Chem 56:7057–7066

Falcone PM, Giudici P (2009) Una nuova prospettiva di studio della qualità degli aceti. Analisi della distribuzione delle dimensioni molecolari Industrie delle Bevande 219:29–37

Falcone PM, Giudici P (2010) Sugar conversion induced by the application of heat to the grape must. J Agric Food Sci 58:8680–8691

Falcone PM, Chillo S, Giudici P, Del Nobile MA (2007) Measuring rheological properties for applications in quality assessment of traditional balsamic vinegar: description and preliminary evaluation of a model. J Food Eng 80:234–240

Falcone PM, Verzelloni E, Tagliazucchi D, Giudici P (2008) A rheological approach to the quantitative assessment of traditional balsamic vinegar quality. J Food Eng 86:433–443

Falcone PM, Boselli E, Frega NG (2011) Structure-composition relationships of the traditional balsamic vinegar close to jamming transition. Food Res Int 44:1613–1619

Falcone PM, Mozzon M, Frega NG (2012) Structure–composition relationships of the traditional balsamic vinegar of Modena close to jamming transition (part II): Threshold control parameters. Food Res Int 45:75–84

Giaccio M, Del Signore A, Cichelli A, Morgante A (1996) Differentiation of traditional balsamic vinegar from common balsamic vinegar by a characteristic compound. Rivista di Merceologia 35:17–28

Giudici P (1993) Acido gluconico: Criterio di genuinità nell'aceto balsamico tradizionale. Industrie delle Bevande 124:123–125

Giudici P, Falcone PM (2010) Zuccheri e Brix, due semplici parametri di discriminazione degli aceti. Industrie delle Bevande 225:7–13

Giudici P, Rinaldi G (2007) A theoretical model to predict the age of traditional balsamic vinegar. J Food Eng 82:121–127

Giudici P, Barbagallo RN, Altieri C, Masini G (1994) Origin evolution of organic acids during the traditional balsamic vinegar aging. Industrie delle Bevande 134:569–574

Lemmetti F (2013) Systemic approach for improving balsamic vinegars. PhD dissertation, University of Modena and Reggio Emilia

Lemmetti F, Giudici P (2010a) Bilancio di massa ed età dell'Aceto Balsamico Tradizionale. Industrie delle Bevande 230:18–28

Lemmetti F, Giudici P (2010b) Gestione della batteria e qualità dell'Aceto Balsamico Tradizionale. Industrie delle Bevande 228:7–16

Lemmetti F, Giudici P (2011) Semplici parametri chimico-fisici per riconoscere gli aceti balsamici invecchiati. Industrie delle Bevande 235:12–24

Lemmetti F, Giudici P (2012) Aceto balsamico tradizionale—manuale di produzione. Edagricole, Bologna

Lemmetti F, Falcone PM, Giudici P (2013) Aceto Balsamico Tradizionale: il ruolo della viscosità sulla qualità oggettiva e sulla qualità percepita. Industrie delle Bevande 244:9–21

Masino F, Chinnici F, Franchini GC et al (2005) A study of the relationships among acidity, sugar and furanic compound concentrations in set of casks for Aceto Balsamico Tradizionale of Reggio Emilia by multivariate techniques. Food Chem 92:673–679

Rubinsztain Y, Yariv S, Ioselis P et al (1986) Characterization of melanoidins by IR spectroscopy—l. Galactose-glycine melanoidins. Org Geochem 9:117–125

Zeppa G, Giordano M, Gerbi V, Meglioli G (2002) Characterization of volatiles compounds in three acetification batteries used for the production of 'Aceto Balsamico Tradizionale di Reggio-Emilia'. Ital J Food Sci 3:247–266

Chapter 7
Sensorial Properties and Evaluation of Balsamic Vinegars

7.1 Official Procedures

Sensory analysis of food is a procedure to assess the qualitative traits that cannot be measured by instrumental analyses, or more importantly, sensory perception completes the chemical and physical characterisation of food. The quality perceived by sensory analysis is the most important trait in food evaluation. The analysis serves several purposes, i.e. to investigate consumer preferences, and to guarantee consistent quality over time.

Official sensory analysis of food must fulfil the guidelines and requirements of the International Organization for Standardization (ISO) and specifically the International Classification for Standards number 67.240 (ICS 2005). For some widely produced foodstuffs, such as PDO wine and olive oil, sensory analysis is compulsory and specified in the corresponding production regulations, but there are no official rules for the sensory analysis of balsamic vinegars, with the exception of TBV.

A number of unofficial quality classification systems exist for BVM, applied exclusively for marketing purposes since their use is limited to a specific brand or manufacturer, without any legal status. In the past, some producers of BVM adopted internal procedures to assess their finished products, including sensory tests. Some have also proposed a standardised sensory test for BVM, to the extent of developing voluntary product certifications audited by external control bodies, but to date none of these systems have been officially recognised. Therefore, they cannot be adopted as official methods, also because they are interpreted as being in conflict with the existing PGI regulations.

The Commission Regulation gives a brief indication of the sensory properties expected for BVM, but no guidelines are provided for testing them (Regulation EC 583/2009). At the time of drafting the present text, sensory analysis of BVM is still performed by individual producers as an internal quality control procedure, in the absence of an official harmonised standard. However, it is likely that in the near

P. Giudici et al., *Balsamic Vinegars*, DOI 10.1007/978-3-319-13758-2_7

future BVM producers will propose an amendment to the PGI regulations, in order to define a common, official method for sensory testing in order to obviate this intrinsic deficiency.

For the reasons expressed above, this chapter will concentrate on the sensory properties and evaluation of TBV, although the observations can be considered equally valid for BVM, since both vinegars are closely related and similar in appearance and chemical–physical traits. The remainder of this chapter is thus applicable to balsamic vinegars in general (BVM and TBV).

7.2 Quality Evaluation of Traditional Balsamic Vinegars

The quality evaluation of TBV is primarily based on sensory analysis, which is also compulsory in order to obtain PDO status. Besides the production process, the product regulations specify the chemical–physical parameters and sensory features for each TBV lot, indicating the compulsory assessment tests performed on a product submitted for authorisation (Gazzetta Ufficiale della Repubblica Italiana 124/2000; Regulation EC 813/2000).

For each batch of TBV, official sensory panels award a final score, which determines the quality and price categories. According to the current assessment procedures applied by the associations of producers, the Consortia of the TBV of Modena (TBVMO), samples can be discriminated into two quality and price levels (*affinato* and *extravecchio*, Fig. 7.1), whereas the Consortia of the TBV of Reggio Emilia (TBVRE) discriminate three quality levels (*aragosta*, *argento*, and *oro*, Fig. 7.2).

Sensory properties determine price, and so evaluation of TBV needs to be as objective and repeatable as possible. The existing official sensory procedures are very similar for the two PDO balsamic vinegars, including appraisal of visual descriptors, followed by olfactory perception and tasting. For TBVRE, the panellists assign scores for the following descriptors: viscosity, colour, and clearness for the visual features; delicacy, straightforwardness, persistence, and acidity for the olfactory features; fullness, maturity, harmony, and acidity for the gustatory features. The panel includes five members from the official list of "ABTRE expert tasters" registered in the Chamber of Commerce of Reggio Emilia.

They individually taste the samples and express their evaluations by compiling a structured questionnaire. Each descriptor has a pre-printed ranking of scores and the panellist chooses the level closest to his evaluation. The use of a structured questionnaire with a pre-printed scale guides the assessments of the panellists, limiting their options for judgement and improving the overall objectiveness of the assessment (Lawless and Malone 1986; Verhagen and Engelen 2006).

Fig. 7.1 Official bottles of Traditional Balsamic Vinegar of Modena (PDO). Two grade levels exist, the red label *affinato* (*refined*) and the gold label *extravecchio* (*extra matured*)

7.2.1 Limitations of the Current TBV Tasting Procedures

Food perception and preferences rely on the ability to combine information from both the taste and olfactory systems. In sensory evaluations, the concept of independent judgement is not realistic in a literal sense, because all the individuals participating in a test have experience as product users, which conditions their responses (Stone et al. 2012).

Food tasting is hindered by some problematic issues, often hidden or difficult to identify. The visual appearance significantly affects the results of the subsequent stages of sensory analysis, and even the olfactory and gustatory sensations greatly influence each other (the halo effect). When designing sensorial assessment procedures, a rational sequence and other devices can minimise the impact of this influence, be it visual, olfactory, gustatory, or trigeminal perception. (Stroebele and De Castro 2004; Auvray and Spence 2008; Spence and Shankar 2010).

In TBV sensory analysis, the tasting procedure consists of a sequence of tests that can significantly alter the final sensory score and the preferences of panellists, because visual appearance is assessed in the first step and this can strongly influence the subsequent sensorial perceptions, irrespective of the real or potential sensory perceptions. The perception of the fragrant and aromatic flavours of TBV is also very problematic because the stimulation of the acetic acid is often so strong that it

Fig. 7.2 Official bottles of Traditional Balsamic Vinegar of Reggio Emilia (PDO). Three grade levels exist, the *aragosta* (*lobster*) label, the *argento* (*silver*) label, and the *oro* (*gold*) label

reduces or completely masks the perception of other volatile substances. All these aspects are crucially important in the sensory analysis of balsamic vinegar and need to be taken into account for a valid assessment.

Studies on TBVRE revealed that the final preference scores were significantly related to the appearance of the samples and, mainly, to their chemical and physical composition. This relationship derives from the common practice of informing panellists of the Brix degree and the acidity values of the samples they taste with the declared intention of helping the panel with the sensory analysis. The assumption is that this information about the TBV assists valid sensory evaluation. In reality, there is good evidence that this information seriously compromises the analysis, conditioning the judgement of the panel members irrespective of their real sensations (Giudici et al. 2009). Furthermore, the significance of these parameters is often confused. For example, the Brix value is often equated with density or used as an indirect index of viscosity or of sugar content (Giudici and Falcone 2010).

It was demonstrated that the scores assigned for individual TBV descriptors are linked because once a panellist has identified his or her favourite vinegar from any particular point of view (taste, visual, or olfactory), he or she tends to favour it for all other sensory attributes, regardless of his or her real sensations. Therefore, an effective and objective sensory analysis is a central issue for TBV production and marketing. In particular it has to satisfy at least two conditions:

- The panellists must be properly trained on the features of TBV.
- The panellists must be as free as possible from any psychological and physical conditioning that might affect human judgement.

7.3 The Balsamic Lexicon

The terminology used in sensory evaluation includes a set of labels that the panel members agree can fully describe the sensory properties of the products being evaluated. Defining the taste and savour of foods and beverages requires clear and unequivocal terms used with the same meaning by all the panellists, with the definitions and terms also being employed to promote the product's characteristics or attributes. The process of developing a product lexicon has four stages:

1. Establish a frame of reference for the product;
2. Establish the reference terms;
3. Review references and examples;
4. Develop a final attribute list.

Numerous lexicons have been developed for a variety of products and procedures including peanut flavour (Johnsen et al. 1987), tactile properties (Civille and Dus 1991), cheddar cheese (Drake et al. 2001), soymilk (N'Kouka et al. 2004), honey (Galán-Soldevilla et al. 2005), and wine (Noble et al. 1987).

In the case of TBV, the first lexicon was developed by Giudici et al. (2009) with a group of 60 panellists who established 24 sensory attributes of traditional balsamic vinegar, referring to four macro-categories (appearance, aroma, taste, and trigeminal sensations). Aroma includes the highest number of attributes (for example, caramel, cooked must, dried prune, honey, apple, liquorice, vanilla, mustard, carob, spices, coffee, and chocolate). These attributes were frequently recognised by all the panellists in the majority of the samples examined. Other frequently used descriptors regard the four consolidated tastes (sweet, sour, bitter, and salty), the tactile perception of viscosity, and the three trigeminal sensations of pungency, spiciness, and astringency. Finally, the current descriptors for TBV have been demonstrated to be redundant and very far from the common sensations of the panellists, making them difficult to quantify and apply.

The lexicon was further expanded by Zeppa et al. (2013) with a set of 17 descriptors for aroma and 3 for taste.

7.4 Towards a New Questionnaire for the Sensory Analysis of Balsamic Vinegars

A standard method for sensory evaluation offers many advantages, as seen for other products with well-established methods, like cheese (Talavera-Bianchi and Chambers 2008), wine (Etaio et al. 2010; Maitre et al. 2010), and dairy products (Coggins et al. 2008). A systematic approach to the sensory evaluation of olive oil is defined and regulated (Regulations EC 2568/91, 796/2002, 1989/2003, 640/2008). However, a standard method for the sensory analysis of vinegars does not yet exist.

Sensory analysis has been used to discriminate vinegar samples on the basis of raw materials (Gerbi et al. 1997, 1998) or processing methods (Tesfaye et al. 2002; Morales et al. 2006). Another study deals with the evaluation of the effects of a tasting protocol on the performance of panellists (Tesfaye et al. 2010).

The design of a good assessment questionnaire for sensory analysis must use clear, unequivocal terms with meanings agreed by all the panellists, while the panellists remain free to assess each attribute as being insufficient, optimal, or excessive. The assessment form needs to be easy to use, so that the panellists can concentrate on their sensory perceptions without being distracted by filling in the questionnaire. Finally, the structure of the questionnaire must not influence the expression of sensory judgement, with numbers or other indications, instead encouraging the use of the full range of the scaled sensory attributes.

A new questionnaire has been designed to satisfy these requirements. It is divided into three sections (Fig. 7.3), each related to one of the three sensory groups (olfactory, gustatory–olfactory, and visual), with respective attributes. The attributes chosen refer to a one-dimensional concept, avoiding ambiguity and ensuring that all the panellists refer to the same sensory concept (ISO 1087/1990, 5492/1992; Giboreau et al. 2007).

The amplitude of each attribute is expressed as an unstructured scale with a straight line of 7 or 14 cm in length, along which the panellist marks his or her degree of perception. The shorter lines are unipolar scales, with the maximum score on the right edge, while the longer lines are bipolar scales, with maximum score in the middle (ISO 4121/1987). Two or three verbal labels are also included for the left, middle, and right limits.

A numeric score is established for the different lines after sensory analysis by overlaying a semi-transparent mask with the different number scales for each descriptor (Fig. 7.4). In the future, the questionnaire could be digitally acquired and the data collected using a computer program (this has not yet been developed). Like the previous evaluation form for TBVRE sensory analysis, the maximum score is 400, but there is no precise correspondence between the scores from the old and new questionnaires.

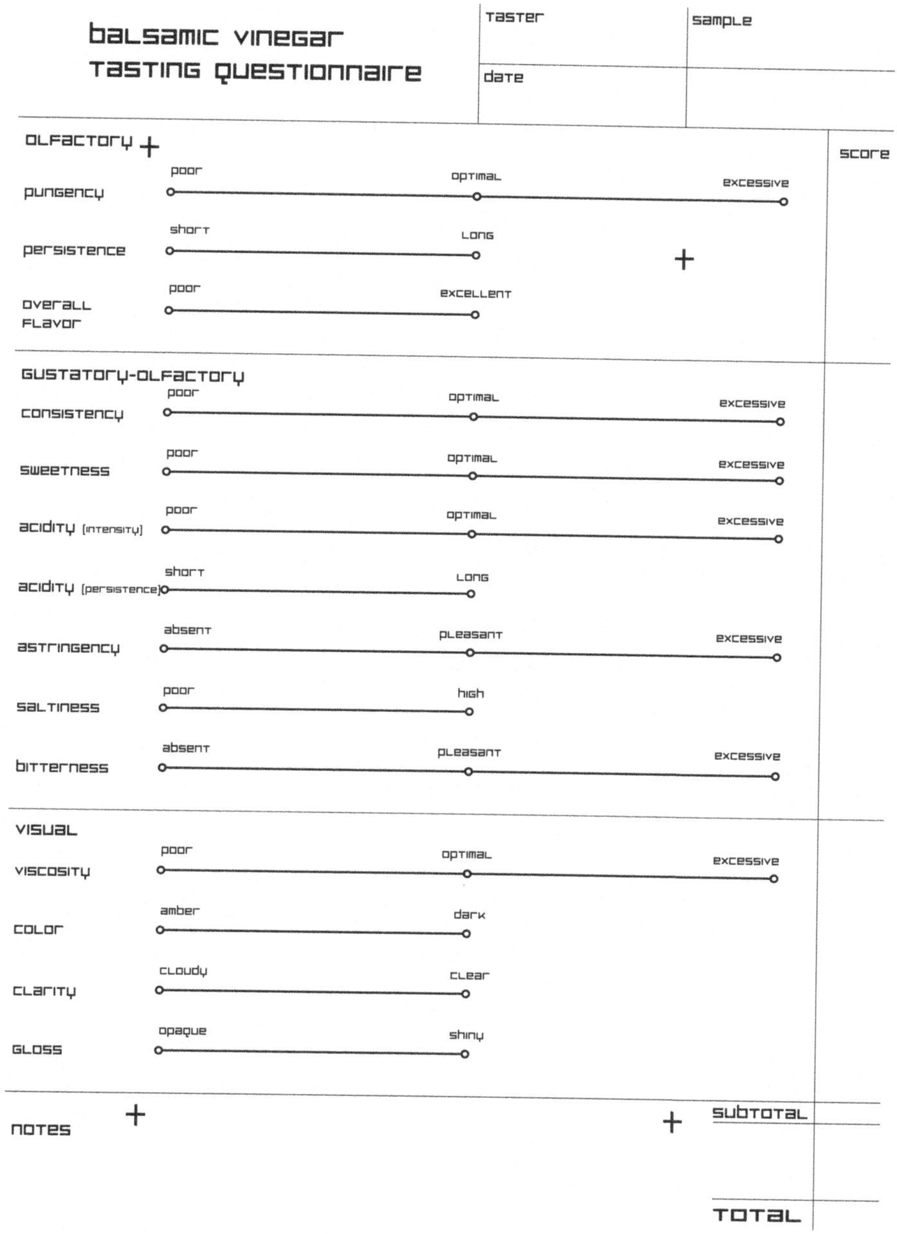

Fig. 7.3 The new proposed questionnaire for the assessment of balsamic vinegars. For each descriptor on the left, a non-numerical interval scale is represented as a continuous line, unipolar or bipolar

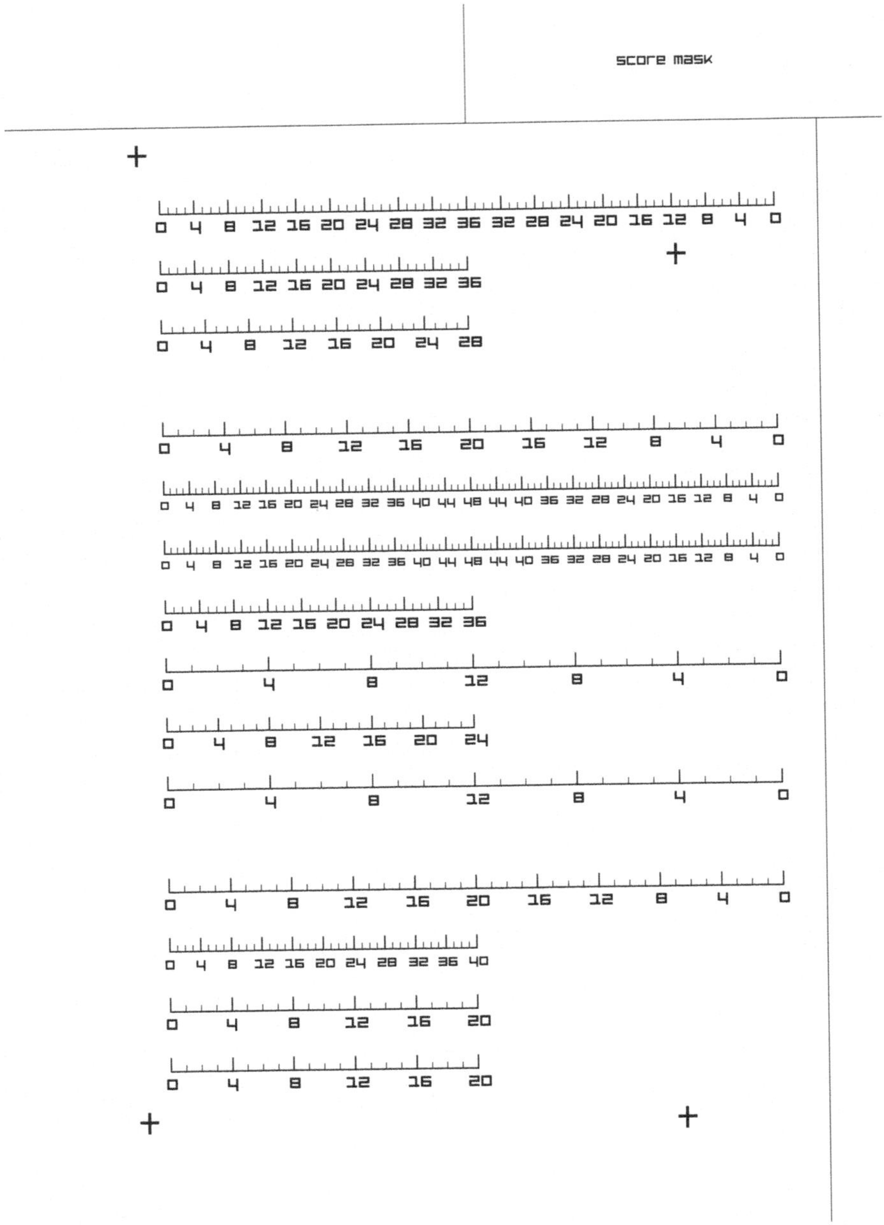

Fig. 7.4 The mask sheet to be overlaid on the questionnaire for the assignment of a numeric score

7.4.1 Tasting Procedure

During the tasting procedure, maximum independence of judgement should be encouraged, and each panellist must feel completely free to express their sensory preferences without the pressure of being judged themselves for their performance. A panellist who is afraid of making mistakes tends to avoid extreme judgements, both very positive and very negative, and tends to favour average values, making their responses totally useless (Lawless and Malone 1986; Meilgaard et al. 1991).

For this reason, individual judgements must remain anonymous for the rest of the panel. Panellists also need to be instructed on how to conduct the sensory analysis. They must abide by their own impressions of each individual attribute in a totally independent way. In this regard, it is very useful to carry out the olfactory analysis using a completely obscured vial or bottle, and complete the relevant section of the questionnaire with no possibility of later correction. The olfactory testing session should be completed before visual assessment, so that interference between the individual sensory attributes is as limited as possible.

Other factors that must be defined for the tasting procedure include temperature, the amount of sample to be tested, the specific gestures to follow for each sensory test, and finally the maximum number of samples to assess in each session.

The panellists may be requested to provide a purely hedonistic comment or to identify a specific level of certain qualitative attributes in relation to a reference standard. In the first case, the panel members do not require specific training because their skill depends on the degree of sensitivity of their sense organs and on the ability to discern different samples based on describable traits. In the second case, training of assessors is mandatory according to the ISO guidelines (ISO 8586/1993, part 1 and 2) to assure the consistency of a panellist's performance and an analytical judgement based on a quality scale defined by assessable standards.

Apart from the visual examination, during tasting the senses involved are smell, taste, and the trigeminal system all working together to determine a global impression of taste, which is then decomposed into the individual aspects of perception. Hybridisation, synergy, and suppression of individual molecules relative to others commonly occur (Frijters 1987; Laffort 1994; Laing et al. 1989).

According to Helson's principle, the sensory effect of a gustatory stimulus of equal extent also varies according to the laws of habituation, sensitisation, and contrast with the level of previous stimulation (Helson 1964).

It is clear that the ability to decompose a complex experience into other less complex ones is a basic requirement for a good panellist, but this also needs to be supported with a simple, fast, and objective evaluation tool. In order to facilitate the isolation of each sensory feature and reduce the mutual influence between them from one sensation to the next, the new questionnaire has been designed to follow the natural sequence of sensory perceptions, which is smell, taste, and visual inspection, in that order.

7.4.2 Olfactory Perceptions

Olfactory sensations are the most difficult to ascertain because the words available to define the large number of flavours are limited. Moreover, other well-known factors, including antagonism, synergy, pseudo-domain, suppression, and adaptation, make it very difficult to identify the aromas in a mixture. In addition, sensory analysis can be complicated by the use of a specific scale for each flavour.

In the case of balsamic vinegars, recognition of aromas is difficult due to the complexity of the mixtures and the maturing process, which tends to degrade the aromatic substances most frequently recognised in recently prepared foodstuffs. Nevertheless, flavour is a key feature for the assessment of balsamic vinegars, and so it is included as an "overall flavour" descriptor, which encompasses the attributes of agreeableness, depth, and complexity.

7.4.3 Gustatory Perceptions

Taste is the least refined of the sensations. It provides an initial survey of the flavour of food (acidity, sweetness, bitterness, and saltiness, without considering the umami and metallic aspects). In order to fully establish a flavour, taste is supported by tactile, thermal, olfactory, and trigeminal perceptions (Murray et al. 2001).

A flavour is always a combination of the four basic descriptors (salty, sweet, sour, and bitter) deriving from soluble substances in the saliva. Even a single chemical substance can produce multiple flavours simultaneously. For example, the salts of sodium and lithium are typically salty, while those of potassium are salty and bitter.

In a similar way, organic acids affect the final taste of balsamic vinegars differently. While tartaric acid is strong and dry, succinic acid is weak and provides bitter and salty notes, and lactic acid is likewise moderately acid. Conversely, malic and citric acids are respectively astringent and freshly sour.

In relation to the metabolites produced by acetic acid bacteria, the acetic acid itself is intense and pungent, while the gluconic acid derived from D-glucose (Giudici 1993) is fresh and sweet. The sodium salts of the same acids have variable effects on salinity, according to the length of the organic chain.

Not all carbohydrates, even if technically called "sugars", have a sweet taste. Some proteins are used as sweeteners, while the majority of peptides are bitter. All these compounds are present in balsamic vinegars in different concentrations, and their balance determines the final taste.

During tasting, the combination of different flavours is not perceived simultaneously. The four flavours have different initial times and persistence in the mouth. Initially sweetness prevails over the others and remains in the mouth approximately until swallowing (a few seconds). After a few seconds, there is a gradual reduction in sweetness and an increase in sour and salty flavours, culminating in acidity and, in particular, bitter aftertastes.

To facilitate the work of decoding a panellist's sensations, the questionnaire considers the intensity of the four flavours, following the same sequence as the sensory taste responses. Moreover, the assessment form includes a quantification of the persistency of the acidity, which generally lasts the longest time after swallowing.

Another parameter affecting sensory perceptions is the temperature of the sample. Cold enhances the freshness of acidity, the dryness of bitterness, and astringency, while it mitigates the sense of softness of sugars (and other substances that give sensations of heat, like alcohols), and the oiliness of substances such as glycerol. In contrast, heat makes acidity and astringency less noticeable and enhances sweetness and texture.

The sensory characteristics of the four tastes are also related to different reactions of the oral mucosa. Sweetness increases the secretion of thick, viscous saliva, whereas acidity and saltiness induce abundant salivary flow. Bitterness usually causes sharply reduced salivation because it is often combined with the sensation of astringency, not arising from the taste buds but from other receptors present on the mucous membranes of the mouth.

7.4.4 Physical Perceptions

Trigeminal tactile sensations derive from the excitement of mechanical and chemical receptors connected to the trigeminal nerve, which is a sensory apparatus distinct from the other senses.

These perceptions can be divided into physical (tactile in the strict sense of the word) and chemical. The former includes the sensations related to consistency, viscosity, grittiness, oiliness, and texture, in addition to the perception of the sample temperature. The latter category includes sensations related to pain, discomfort, and irritation such as astringency, pungency, spiciness, sizzling, metallic, pseudo-heat, and pseudo-freshness (Prescott and Stevenson 1995).

For balsamic vinegars, the most relevant attributes in this group are definitely pungency, viscosity (consistency), and astringency. The viscosity of TBV plays an important role in the sensorial perception of quality, as already demonstrated for other beverages. Viscosity modifies other oral sensations in wine, including saltiness, sweetness, bitterness, flavour, and astringency (Christensen 1980; Hollowood et al. 2002; Wendin et al. 1997).

Furthermore, in beer and wine the molecules that stimulate the sensation of sweetness are the same as those that contribute to the "body" or "oiliness" because of their ability to increase the viscosity.

The perceived viscosity of liquid and semisolid foods is usually assessed either by oral methods based on the measurement of resistance to flow over the tongue surface or by non-oral methods based on the measurement of the force required to manipulate the fluid in the mouth while sipping from a teaspoon (Susan et al. 1991). Moreover, due to the subjectivity of human perception and to the time required to train panel members, sensory analysis may be poorly reproducible and very expensive.

Recently, an accurate instrumental method to characterise the rheological properties and the viscosity of TBV was set up; the results indicate a wide range of mechanical responses, from Newtonian to pseudo-plastic behaviour. The instrumental determination of viscosity was related to sensory analysis of TBV (Falcone et al. 2008). The results were confirmed in the collaborative study between the Unimore Culture Collection (http://www.umcc.unimore.it/) and the sensory panel test of the "*Confraternita* of TBV of Reggio Emilia". In particular, they studied the viscosity of the top 20 sensory samples and 60 chosen randomly from among all the samples participating in the competition (Lemmetti 2013). The authors reported that the sensory panel generally prefered TBV samples with high viscosity (Fig. 7.5).

The viscosity of TBV is a variable influenced by the apparent density (specific gravity) and the nature of the chemical compounds. In general, mature TBV is more viscous than less matured samples. In the specific case reported here, the top sensory samples (F) are generally more viscous than the other samples with the same specific gravity (Fig. 7.6); which could be due to the presence of high-molecular weight compounds such as melanoidins (see Chap. 6). Furthermore, matured samples behave differently from a rheological point of view (Falcone and Giudici 2008; Lemmetti et al. 2013).

7.4.5 Visual Appearance

Visual perceptions are very important factors for judgements, and visual assessments should be conducted as the last stage of evaluation to avoid influencing the other analyses.

The colour of TBV ranges from amber to dark brown and it is usually indicative of the age of the vinegar (Falcone and Giudici 2008). Other visual attributes are brilliance and clearness. Brilliance refers to the glossy surface of a drop of vinegar and is a measure of reflectivity, producing an effect similar to polished metal or a drop of mercury, although transparent. Many promotional images of balsamic vinegars play on this brilliance, for example, by showing it as the vinegar sprinkles down in drops. Clearness is a measure of transparency and is assessed by observing a thin layer of balsamic vinegar through a transparent surface such as glass. The amount of matter in suspension and the size of the particulates may affect clearness, making the vinegar opalescent or turbid. Finally, viscosity, which is also perceived during the gustatory test, refers to the thickness and persistence of a layer of sample sliding along the side of a flask.

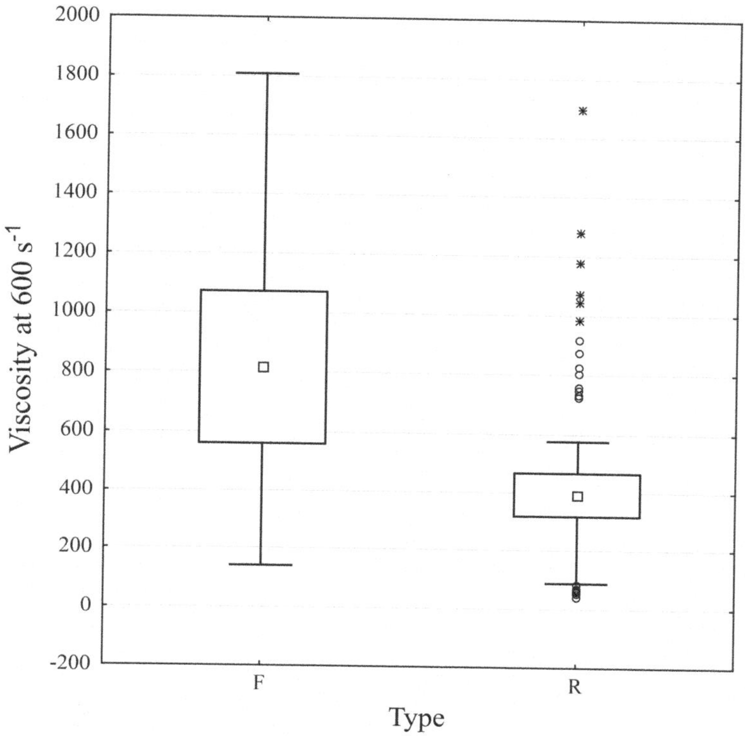

Fig. 7.5 Viscosity at a shear rate of 600 s^{-1} of the best quality samples (F) and random samples (R). F and R samples come from the competition *Palio Matildico* 2010. *Box* and *whisker plots* show mean, mean$\pm$0.95 confidence interval, non-outliers range, outliers, and extremes (adapted from Lemmetti 2013)

7.4.6 Guidelines for Sensory Score Ranking

Guidelines are provided below for the implementation of each sensory test, together with a description of the characteristics and the corresponding maximum scores, expressed as a percentage of the overall evaluation. For quick and concise definitions of the terms commonly used in sensory analysis, see Sect. 7.5.

The olfactory test is performed before seeing the sample, and the vinegar is kept in a darkened flask during the test. The three descriptors, accounting for up to 25 % of the total score, are pungency (9 %), persistence (9 %), and overall olfactory flavours (7 %). The last is a judgement of pleasantness that takes into account the presence of more or less pronounced characteristic aromas of balsamic vinegars. Pungency is the first example of a pyramidal (bipolar) scale descriptor; the score ranges from zero (poor) to 36 (optimal) and then decreases again to zero (excessive).

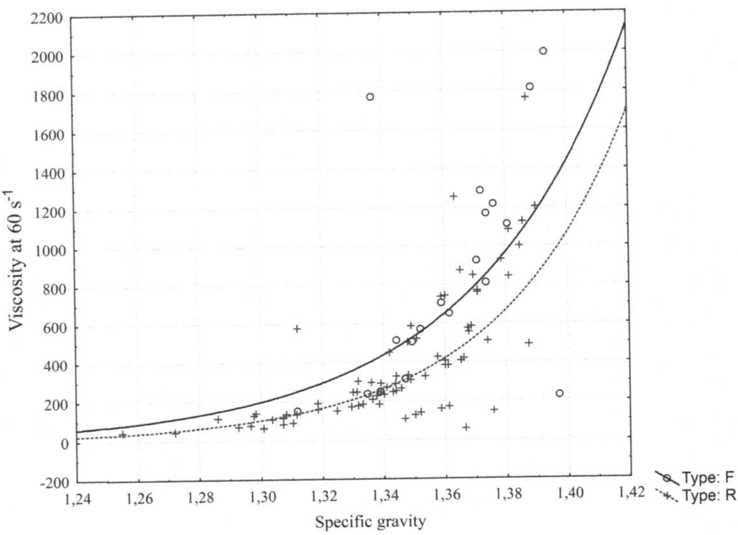

Fig. 7.6 Viscosity vs. specific gravity of top samples (*circle marker*) and random samples (*cross marker*). Top and random samples come from the competition *Palio Matildico* 2010 (adapted from Lemmetti et al. 2013)

After the olfactory test, the gustatory–olfactory test is performed, still keeping the sample obscured in a darkened flask. The partial score represents up to 50 % of the total score, and the descriptors are consistency (5 %), which expresses the tactile impression in the mouth mainly due to density and viscosity, sweetness (12 %), intensity of acidity (12 %), persistence of acidity (9 %), astringency (3 %), saltiness (6 %), and bitterness (3 %). With the exception of persistence and saltiness, which obviously cannot be excessive, all the descriptors have pyramidal scales, as for pungency.

The visual test score represents up to 25 % of the total. This test assesses viscosity (5 %), which is the ability of the liquid to adhere to the side of a flask, colour and the presence of undesirable reflections (10 %), clearness (5 %), referred to the transparency of the sample, and gloss (5 %), which is the ability to reflect light. Viscosity is the only visual test evaluated on a pyramidal scale, penalising both excessively viscous and excessively fluid samples.

7.4.7 Conclusions

Sensory analysis is a very powerful and useful tool, offering significant contributions to the quality assessment of TBV, in combination with chemical and instrumental analyses.

An essential prerequisite for the effectiveness and reliability of TBV sensory analysis is the establishment of procedures that maximise the independence of judgement of the panellists. Precise tasting procedures and clear evaluation questionnaires both help to ensure a reasonable degree of independence during sensory evaluation. The evaluation form proposed here has an easy-to-use layout, differing from those currently used by the TBVMO and TBVRE *Consortia*, but fulfilling the most recent directives regarding sensory analysis and formulated on a scientific basis. It applies clear, easily identifiable, and officially recognised terms (i.e. bitter, sweet, salty, astringent, and pungent) that have clear meanings shared by all panel members.

Furthermore, the questionnaire avoids the use of categorical values in order to encourage the use of the complete scale. This simple layout enables panellists to concentrate more on their perceptions and less on the compilation of the questionnaire.

The future validation of this and other standardised sensory methods will help to obviate the drawbacks in the current TBV evaluation systems, offering a more reliable and rational classification of TBV samples into quality and price categories.

7.5 Vocabulary

Acidity Sourness is due to the presence of undissociated organic acids, which convey a feeling of acidity varying in intensity and often associated with other sensory perceptions (taste and/or trigeminal sensations). The intensity of the acidity is the level of maximum stimulation achieved during the taste, while the persistence quantifies the temporal duration of the acid stimulus. A sour taste is usually intensified at low temperatures. In balsamic vinegars, the acid taste comes from the influence of all the acid species present in the product that contribute to the fixed acidity (malic, tartaric, succinic, citric, gluconic, and lactic acid) and volatile acidity (acetic acid) also responsible for the pungency.

Adaptation Adaptation is the reduction or interruption of the transmission of the perception signal by the sensory receptors, when they are subjected to a stimulus that persists for too long and with the same intensity.

Addiction Addiction is similar to adaptation: the sensory receptors develop certain insensitivity to flavours after prolonged exposure, so the intensity of the receptor response is reduced according to the duration of the stimulus. Therefore, in the olfactory test of sensory analysis, appropriate breaks must be inserted between samples.

Antagonism Antagonism originates from the competition between different molecules affecting the same sensory receptor. The overall perception depends on the bond strength and the concentration of the species that form the agonist–antagonist pair.

Aroma Aroma usually refers to pleasant smells, as a synonym of fragrance, scent; it is the set of olfactory sensory perceptions triggered by gas molecules that reach the olfactory epithelium. The perception of flavour is generally enhanced by temperature increase.

Astringency Astringency is the simultaneous feeling of freshness and dryness as a result of a trigeminal type stimulation. Astringency is due to the presence of substances that cause the aggregation and precipitation of the proteins of the saliva, resulting in a decrease in oral lubrication. Astringency is often combined with a bitter taste due to the molecules responsible for both effects. Like bitterness, astringency is intensified by low temperatures.

Bitterness Bitterness is usually an aftertaste and it is intensified at low temperatures. If moderate, bitterness can be a pleasant character in balsamic vinegars. The bitterness of balsamic vinegar comes from several natural organic substances, from the grapes and the wood of the barrels, as well as from the technological processes, *e.g.* cooking and microbiological processes.

Brilliance Brilliance refers to the surface gloss, i.e. the capacity to reflect light, creating the effect of lucidity, similar to a polished metal surface. Many pictures of commercial balsamic vinegar play on brilliance as an attractive attribute for marketing purposes.

Clearness Clearness is an attribute for the visual assessment of the transparency of a sample. Looking at a thin layer of balsamic vinegar through a transparent surface, such as glass, the presence of fine material in suspension may affect the clearness and turn the sample opalescent or turbid, depending on the size of the particulates.

Colour Colour can be precisely defined and measured, but only with very expensive equipment. For the evaluation of the colour of a food by a panel of tasters, details of the procedure have to be defined, such as the intensity and temperature of the source light.

Consistency This attribute, during the gustatory–olfactory test, refers to the overall tactile sensation perceived as soon as the sample is put into the mouth. The consistency refers to the syrupy effect due to the viscosity and specific weight, thus related to the dry extract and the presence of high-molecular weight substances responsible for the rheological behaviour. Oiliness and granularity also contribute to the consistency of a sample. Relative high temperatures accentuate the perception of consistency, which is indeed distinctly felt as soon as the sample is heated on the tongue.

Density (relative) Relative density, also specific weight, in physical terms is the ratio between the density of a substance and water at the same temperature. In balsamic vinegar, it depends on the nature and amount of substances other than water. Density should not be confused with viscosity as they are two different parameters: density is the amount of matter per volume unit; viscosity is a measurement of the cohesive force of a fluid.

Flavour Flavour refers to the total integrated response of all the sense organs involved in tasting (gustatory, olfactory, and trigeminal).

Hybridisation Hybridisation originates from the olfactory perception of a mixture of substances, which generates a smell different from that of the components.

Persistence Persistence stands for the duration of the sensory stimulation. As regards the olfactory test, persistence indicates whether the perception of pungency and aroma lasts for a relatively short or long time after the interruption of the stimulation. In the gustatory–olfactory test, the persistence descriptor is coupled with the intensity to precisely define the type of acidity of the sample.

Pseudo-heat and pseudo-freshness These sensations are trigeminal perceptions of temperature different from the actual temperature of the sample, due to the action of specific classes of molecules, *e.g.* alcohols (pseudo-heat) and some acids (pseudo-freshness).

Pungency The pungency is the simultaneous feeling of irritation, pain, and burning, which are trigeminal stimulations felt during the olfactory test. In balsamic vinegar, pungency is due to acetic acid in gaseous state, and it is thus closely related to volatile acidity.

Saltiness Saltiness is generally perceived more distinctly at the end of tasting, when sweet and sour sensations fade. A strong persistence of salivation after swallowing is a sign of high salinity. The perception of saltiness is due to the presence of cations (mainly sodium and potassium) of the salified acids.

Sensitisation Sensitisation is the opposite phenomenon to addiction. The intensity of the receptor response, and therefore of the perception, remains high even if the intensity of the stimulus decreases, until adaptation prevails once more.

Spiciness Spiciness is a trigeminal perception of simultaneous heat and irritation.

Synergy Synergy originates from the sensory perception of a mixture of substances, which generates a stimulus more intense than the sum of the *stimuli* of the single components.

Suppression Suppression originates from the sensory perception of a mixture of two or more substances, which generates the annulment of the stimulus caused by a single substance (unidirectional suppression) or both (two-way suppression).

Sweetness In balsamic vinegar, sweetness is the first taste perceived during the gustatory test. The sweet taste initially tends to mitigate the acidity. Sweetness contributes strongly to the feeling of fullness, softness, and texture of the sample, and it is mitigated by low temperatures.

Trigeminal senses In addition to the olfactory and the gustatory senses, the sensory system connected to the trigeminal nerve is sensitive to tactile/mechanical stimulations (consistency, viscosity, lubricity, grittiness, temperature, etc.) as well as to other chemical-induced perceptions (astringency, pungency, spiciness, pseudo-heat, pseudo-freshness, etc.).

Viscosity Viscosity is a measure of the strength of the molecules of a fluid in oppose flow, and it is easily measured instrumentally. More specifically, viscosity is the resistance of a fluid to gradual deformation by shear stress or tensile stress. Viscosity is therefore influenced by the presence of substances able to increase the intermolecular forces, generally with high molecular weight. There is no physical relationship between viscosity and density, although high-density balsamic samples are often the most viscous. The viscosity, perceived during the gustatory–olfactory test, is included under the "consistency" descriptor. In visual evaluation, viscosity refers to the thickness and persistence of a layer of sample sliding along the side of a flask.

References

Auvray M, Spence C (2008) The multisensory perception of flavour. Conscious Cogn 17:1016–1031

Christensen CM (1980) Effects of taste quality and intensity on oral perception of viscosity. Percept Psychophys 28:315–327

Civille GV, Dus CA (1991) Evaluating tactile properties of skin care products: a descriptive analysis technique. Cosmet Toiletries 106:83–88

Coggins PC, Schilling MW, Kumari S, Gerrard PD (2008) Development of a sensory lexicon for conventional milk yogurt in the United States. J Sens Stud 23:671–687

Drake MA, Mcingvale SC, Gerard PD et al (2001) Development of a descriptive language for cheddar cheese. J Food Sci 66:1422–1427

Etaio I, Albisu M, Ojeda M et al (2010) Sensory quality control for food certification: a case study on wine. Method development. Food Control 21:533–541

Falcone PM, Giudici P (2008) Molecular size and molecular weight distribution impacting Traditional Balsamic Vinegar ageing. J Agr Food Chem 56:7057–7066

Falcone PM, Verzelloni E, Tagliazucchi D, Giudici P (2008) A rheological approach to the quantitative assessment of Traditional Balsamic Vinegar quality. J Food Eng 86:433–443

Frijters JE (1987) Psychophysical models for mixtures of tastants and mixtures of odorants. Ann N Y Acad Sci 510:67–78

Galán-Soldevilla H, Ruiz Pérez-Cacho P, Jiménez S et al (2005) Development of a preliminary sensory lexicon for floral honey. Food Qual Prefer 16:71–77

Gazzetta Ufficiale della Repubblica Italiana n. 124; 2000—Disciplinare di produzione della denominazione di origine protetta "Aceto Balsamico Tradizionale di Reggio Emilia"

Gerbi V, Zeppa G, Antonelli A, Carnacini A (1997) Sensory characterization of wine vinegars. Food Qual Prefer 8:27–34

Gerbi V, Zeppa G, Beltramo R, Carnacini A, Antonelli A (1998) Characterization of white vinegars of different sources with artificial neural networks. J Sci Food Agr 78:417–422

Giboreau A, Dacremont C, Egoroff C et al (2007) Defining sensory descriptors: towards writing guidelines based on terminology. Food Qual Prefer 18:265–274

Giudici P (1993) Gluconic acid: genuine criterion of the traditional balsamic vinegar. Industrie delle Bevande 124:123–125

Giudici P, Falcone PM (2010) Zuccheri e Brix, due semplici parametri di discriminazione degli aceti. Industrie delle Bevande 35:7–23

Giudici P, Falcone PM, Scacco A, Lanza CM (2009) Sensory analysis of the traditional balsamic vinegar. Industrie delle Bevande 224:27–42

Helson H (1964) Adaptation-level theory. Harper & Row, Oxford, UK

Hollowood TA, Linforth RST, Taulor AJ (2002) The effect of viscosity on the perception of flavour. Chem Senses 27:583–589

International Classification for Standardization (ISO) (2005) 6th edn

International Organization for Standardization (ISO) 5492/1992. Sensory analysis. Vocabulary

International Organization for Standardization (ISO) 4121/1987. Sensory analysis. Methodology. Evaluation of food products by methods using scales

International Organization for Standardization (ISO) 8586/1993. Sensory analyses. General guidance for the selection, training and monitoring of assessors. Part 1: Selected assessors

International Organization for Standardization (ISO) 8586/1993. Sensory analysis. General guidance for the selection, training and monitoring of assessors. Part 2: Experts

International Organization for Standardization (ISO) 1087/1990. Sensory analysis. Terminology

Johnsen PB, Civille GV, Vercellotti JR et al (1987) Development of a lexicon for the description of peanut flavor. J Sens Stud 3:9–17

Laffort P (1994) An interaction model for odor quality and intensity. Percept Psycophys 55:363–372

Laing DG, Cain WS, McBride RL, Ache BW (1989) Perception of complex smells and tastes. Academic, Sidney

Lawless H, Malone G (1986) The discriminative efficiency of common scaling methods. J Sens Stud 1:85–98

Lemmetti F (2013) Systemic approach for improving balsamic vinegars. PhD dissertation, University of Modena and Reggio Emilia

Lemmetti F, Falcone PM, Giudici P (2013) Traditional balsamic vinegar: the role of the viscosity on the objective and perceived quality. Industrie delle Bevande 244:9–21

Maitre I, Symoneaux R, Jourjon F, Mehinagic E (2010) Sensory typicality of wines: how scientists have recently dealt with this subject? Food Qual Prefer 21:726–731

Meilgaard M, Civillie GV, Carr BT (1991) Sensory evaluation technique, 4th edn. CRC, Boca Raton, FL

Morales ML, Benitez B, Tesfaye W et al (2006) Sensory evaluation of Sherry vinegar: traditional compared to accelerated aging with oak chips. J Food Sci 71:238–242

Murray JM, Delahunty CM, Baxter IA (2001) Descriptive sensory analysis: past, present and future. Food Res Int 34:461–471

N'Kouka KD, Klein BP, Lee SY (2004) Developing a lexicon for descriptive analysis of soymilks. J Food Sci 69:259–263

Noble AC, Arnold RA, Buechsenstein J et al (1987) Modification of a standardized system of wine aroma terminology. Am J Enol Viticult 38:143–146
Prescott J, Stevenson RJ (1995) Pungency in food perception and preference. Food Rev Int 11:665–698
Regulation (EC) No 2568/91 on the characteristics of olive oil and olive residue oil and on the relevant methods of analysis. Off J Eur Communities 1991(248):1–83
Regulation (EC) 1989/2003 amending Regulation No 2568/91/EEC. Off J Eur Communities 2003 (295):57–77
Regulation (EC) 640/2008 amending Regulation No 2568/91/EEC. Off J Eur Communities 2008 (178):11–16
Regulation (EC) No 583/2009 entering a name in the register of protected designations of origin and protected geographical indications [Aceto Balsamico di Modena (PGI)]. Off J Eur Communities 2009(52):7–11
Regulation (EC) No 796/2002 amending Regulation No 2568/91/EEC. Off J Eur Communities 2002(128):8–28
Regulation (EC) No 813/2000 supplementing the Annex to Commission Regulation (EC) No 1107/96 on the registration of geographical indications and designations of origin under the procedure laid down in Article 17 of Regulation (EEC) No 2081/92. Off J Eur Communities 2000(43):5–6
Spence C, Shankar MU (2010) The influence of auditory cues on the perception of, and responses to, food and drink. J Sens Stud 25:406–430
Stone H, Bleibaum R, Thomas HA (2012) Test strategy and the design of experiments. In: Stone H, Bleibaum R, Thomas HA (eds) Sensory evaluation practices, 4th edn. Academic, London
Stroebele N, De Castro JM (2004) Effect of ambience on food intake and food choice. Nutrition 20:821–838
Susan A, Langstaff SA, Guinard JX, Lewis MJ (1991) Instrumental evaluation of the mouthfeel of beer and correlation with sensory evaluation. J Brewing Inst 97:427–433
Talavera-Bianchi M, Chambers DH (2008) Simplified lexicon to describe flavor characteristics of western European cheeses. J Sens Stud 23:468–484
Tesfaye W, Morales ML, García-Parrilla MC, Troncoso AM (2002) Wine vinegar: technology, authenticity and quality evaluation. Trends Food Sci Technol 13:12–21
Tesfaye W, Morales ML, Callejón RM et al (2010) Descriptive sensory analysis of wine vinegar: tasting procedure and reliability of new attributes. J Sens Stud 25:216–230
Verhagen JV, Engelen L (2006) The neurocognitive bases of human multimodal food perception: sensory integration. Neurosci Biobehav 30:613–650
Wendin K, Solheim R, Allmere T, Johanson L (1997) Flavour and texture in sour milk affected by thickeners and fat content. Food Qual Prefer 4:281–287
Zeppa G, Gambigliani Zoccoli M, Nasi E et al (2013) Descriptive sensory analysis of Aceto Balsamico Tradizionale di Modena DOP and Aceto Balsamico Tradizionale di Reggio Emilia DOP. J Sci Food Agr 93:3737–3742

Chapter 8
"Balsamic": A Global Debate

8.1 The Issue of European Protected Denominations on the Global Market

The World Trade Organization (WTO) protects a form of intellectual property known as "Geographical Indications" (GI). Under the agreement on Trade-Related Aspects of Intellectual Property Rights (TRIPS), GI's are "indications which identify a product as originating in the territory of a Member, or a region or locality in that territory, where a given quality, reputation or other characteristic of the product is essentially attributable to its geographical origin" (TRIPS text by WTO).

The TRIPS agreement mandates two levels of protection for GI. On the first level, GI for any goods provides a form of consumer protection and prevents unfair competition. The GI for wines and spirits are relatively new geographical appellations, and their level of protection is higher than for food. In fact, there is no need to demonstrate that consumers might be misled or that the use constitutes an act of unfair competition. Well-known examples of GI include *Champagne*, *Tequila*, and *Roquefort*. GIs are "commercially valuable source identifiers" and, thus, are similar to trademarks in terms of their function and value. GI and trademarks are also similar in terms of the rationale behind their legal protection, which is consumer protection and prevention of unfair competition. Misuse of GI's can be detrimental to consumers' interests and constitute unfair competition between producers (Rovamo 2006).

The status of protected GI's is a legal structure defined by the European Union to protect regional food and specialities. It groups together three distinctive systems, Protected Designation of Origin (PDO), Protected Geographical Indication (PGI), and Traditional Speciality Guaranteed (TSG) (Regulation EU 1151/2012).

There are also separate regimes for spirits and aromatised drinks (geographical designations) as well as for wines (geographical indications or appellations). Moreover, the latter rules are dependent on the country in which the wine is produced.

P. Giudici et al., *Balsamic Vinegars*, DOI 10.1007/978-3-319-13758-2_8

Fig. 8.1 The three logos for foodstuff with protected geographic indication status

The official community symbols for labelling or packaging of products with registered PDO, PGI, or TSG names are shown in Fig. 8.1. The original and currently valid logos are reported in the annexes of Regulation EC 628/2008 and Regulation EC 1216/2007.

The use of geographical appellations to distinguish food and beverages, especially wine, is very old and dates back to ancient times. It was applied systematically for the first time in 1935 for French wines, with the constitution of the *Institut National des Appellations d'Origine*. Geographical appellation became legally enforceable with the constitution of the *Appellation d'origine contrôlée*.

Today, geographical appellations or GI's are quite widespread and many countries have adopted systems that resemble the GI structure of the EU, where the number of regional foods with GI status is huge. France claims 170 PDO/PGI and 640 wines with appellation; Italy has 193 PDO/PGI products and 467 wines with appellation. The other countries follow with relatively less products (DG Agriculture and Rural Development Working document 2008). These large numbers clearly indicate how seriously the EU considers GI status. This is due to the general idea that GI status is a key for improving the social and economic prosperity of local agricultural areas.

GI's are credited with offering five main benefits (O'Connor and Company 2005):

- Promoting products with specific characteristics, particularly those coming from less-fortunate or rural areas
- Promoting rural and agricultural activity in general
- Improving the income of farmers, in return for a "genuine effort to improve quality"
- Retention of population in rural areas
- Helping consumers discriminate between authentic and non-genuine products, through clear and succinct information regarding product origin.

The origin of a product is only one of the criteria necessary in order to qualify for protected GI, and it must also meet various quality criteria, strictly or exclusively linked to the geographical environment, including natural and human factors. The definition and criteria for PDO, PGI, and TGS status are listed below.

PDO refers to a specific place or, in exceptional cases, to the name of a country, used as a designation for an agricultural product or a foodstuff:

- Which comes from such an area, place, or country,
- Whose quality or properties are significantly or exclusively determined by the geographical environment, including natural and human factors,
- Whose production, processing, and preparation take place within the defined geographical area.

PGI is the name of an area, a specific place, or, in exceptional cases, the name of a country, used as a description of an agricultural product or a foodstuff:

- Which comes from such an area, place, or country,
- Which has a specific quality, goodwill, or other characteristic property, attributable to its geographical origin,
- Whose production, processing, or preparation takes place within the defined geographical area.

PSG is a trademark for an agricultural product or a foodstuff, which has a certain feature or a set of features, setting it clearly apart from other similar products or foodstuff belonging to the same category. The product or foodstuffs must be manufactured using traditional ingredients or must be characteristic for its traditional composition, production process, or processing, reflecting a traditional type of manufacturing or processing.

The European system of GI is not restricted to EU products, and several products from outside Europe have PDO and PGI status, "Zhenjiang Xiang Cu" vinegar being a relevant example (see Sect. 1.1).

8.2 National and International Acceptance of Protected Balsamic Vinegars

Currently there are four vinegars with PDO and two with PGI status. The first vinegars to achieve GI status were the two Traditional Balsamic Vinegars of Modena and Reggio Emilia in the year 2000. The most recent was the PGI for Zhenjiang Xiang Cu in the year 2012.

A third balsamic vinegar, Aceto Balsamico di Modena, was awarded PGI status in 2009 (Regulation EC 583/2009). Some member states, in particular Germany and Greece, objected not to the entire name but to the words *aceto*, *balsamico*, and *aceto balsamico*, or translations thereof. Ultimately, PGI grants protection to the name *Aceto Balsamico di Modena* as a whole. Individual non-geographical components of the name can be used, even jointly, and also in translation.

The Consortium for the Protection of Balsamic Vinegar of Modena PGI is critical of the fact that protection is granted only for the entire name and not for the individual terms and is thus in favour of legal action against producers labelling vinegars or condiments with the term "balsamic". The position of the Consortium is surprising for at least two reasons. The first is that the EU regulation clearly states that protection refers to the whole name and not to the individual terms. The second somewhat amusing observation is that important members of the Consortium are also members of The Vinegar Institute, where the term "balsamic" is used freely for several kinds of balsamic vinegars, for example:

> Balsamic vinegars produced domestically in the United States (U.S.) and North America are made from wine vinegar blended with grape juice or grape "must" and caramel may be added in small amounts for color stability. Some juice may be subjected to an alcoholic and subsequent acetous fermentation and some to concentration or heating. These products typically have a clean balsamic vinegar flavor and aroma with a sweet and sour taste. The color is typically dark brown, except for white balsamic vinegars. (Source: The Vinegar Institute website).

To date the Consortium's stance has not affected the labelling of balsamic products around the world and even in EU member states, but many small producers of condiments and sauces, who label their products with the word balsamic, had some legal disturbance. In the opinion of the authors, supported by the sentences of the Civil Tribunals, the word *balsamic* is a generic term.

Article 18 of EU regulation No 1151/2012 clearly states one of the objectives of protected GI as being:

> ... providing clear information on the products with specific characteristics linked to geographical origin, thereby enabling consumers to make more informed purchasing choices.

In the specific case of the protected GI of the three balsamic vinegars, *Aceto Balsamico Tradizionale di Modena*, *Aceto Balsamico di Modena*, and *Aceto Balsamico Tradizionale di Reggio Emilia*, is questionable whether consumers are really enabled to make more informed purchasing decisions. This is in reality a rhetorical question because the answer is obviously negative. A balsamic PGI shares, respectively, three terms with the Modena PDO and two with the Reggio Emilia PDO, but in reality the PGI product is made from a completely different procedure and, furthermore, the must and vinegar do not necessarily come from local grapes (see Chap. 4). Clearly, this can hardly help clarify the differences between balsamic products with protected GI in the minds of consumers.

8.3 Problems Related to Protected Geographical Status in the EU

Today, with a few decades of experience since the implementation of GI, it is time to draw up a balance of the costs and benefits of the system. The first question to pose is whether the protected GI system achieved its original aims as listed above. In particular, does it help and promote rural areas, and are producers able to obtain better prices for authentic products? Considering that many PDO products suffer from cyclical commercial problems and that market prices are often lower than production costs, then the answer has to be "no". Furthermore, the prices paid by consumers for PDO products are higher than those for analogous products, which implies that something has gone seriously wrong in the distribution chain.

References

DG Agriculture and Rural Development Working document (2008) http://ec.europa.eu/agriculture/quality/policy/workingdocs/gi_en.pdf. Accessed May 2014

O'Connor and Company (2005) Geographical indications and the challenges for ACP countries. A discussion paper. http://agritrade.cta.int/en/content/view/full/1794. Accessed May 2014

Regulation (EC) 1216/2007 laying down detailed rules for the implementation of Council Regulation (EC) 509/2006 on agricultural products and foodstuffs as traditional specialties guaranteed

Regulation (EC) 628/2008 amending Regulation (EC) 1898/2006 laying down detailed rules of implementation of Council Regulation (EC) 510/2006 on the protection of geographical indications and designations of origin for agricultural products and foodstuffs

Regulation (EC) 583/2009 entering a name in the register of protected designations of origin and protected geographical indications Aceto Balsamico di Modena (PGI)

Regulation (EU) 1151/2012 of the European Parliament and Council on quality schemes for agricultural products and foodstuffs

Rovamo O (2006) Monopolising names? The protection of geographical indications in the European Community. Helsinki University Faculty of Law

TRIPS on WTO website. http://www.wto.org/english/docs_e/legal_e/27-trips_01_e.htm. Accessed January 2014

The Vinegar Institute. http://www.versatilevinegar.org/todaysvinegar.html. Accessed May 2014

The manufacturer's authorised representative in the EU is Springer Nature Customer Service Centre GmbH, Europaplatz 3, 69115 Heidelberg, Germany. If you have any concerns regarding our products, please contact ProductSafety@springernature.com

Printed and bound by CPI Group (UK) Ltd, Croydon, CR0 4YY
25/09/2026
02222836-0019